Canine Genetics, Health and Medicine

Edited by Catrin Rutland

Published in London, United Kingdom

IntechOpen

Supporting open minds since 2005

Canine Genetics, Health and Medicine
http://dx.doi.org/10.5772/intechopen.92513
Edited by Catrin Rutland

Part of IntechOpen Book Series: Veterinary Medicine and Science, Volume 7
Book Series Editor: Rita Payan-Carreira
Contributors
Fatimaezzahra Amarir, Abdelmohcine Aimrane, Abdelkbir Rhalem, Catrin Rutland, Siobhan Simpson, Kara-Zoë Kordtomeikel, Serena Wong, Samantha Bennison, Samir El-Gendy, Malcolm Cobb, Andreia A.F. Santos, Tatiane Moreno Ferrarias Epiphanio, Edo D' Agaro, Andrea Favaro, Davide Rosa, Claudete Paula, Mario Bonci, Bianca Navarro, Diana Nascimento, Miriam Rodríguez Esquivel, Héctor Guerrero Flores, Alejandro Sánchez, Emmanuel Salcedo, Teresa Apresa García, Ricardo López Romero, Florinda Jiménez Vega, Rodrigo Aldariz, Mauricio Salcedo, Alejandra Valdivia Flores, Lung-Kwang Pan, Chih-Feng Chen, Shao-Wen Chiu, Chao-Hsun Chuang, Verónica Flores, Luciano Ritossa, Gustavo Viozzi

First published in London, United Kingdom, 2021 by IntechOpen
IntechOpen is the global imprint of INTECHOPEN LIMITED, registered in England and Wales, registration number: 11086078, 5 Princes Gate Court, London, SW7 2QJ, United Kingdom
Printed in Croatia

British Library Cataloguing-in-Publication Data
A catalogue record for this book is available from the British Library

Additional hard and PDF copies can be obtained from orders@intechopen.com

Canine Genetics, Health and Medicine
Edited by Catrin Rutland
p. cm.
Print ISBN 978-1-83962-654-8
Online ISBN 978-1-83962-655-5
eBook (PDF) ISBN 978-1-83962-656-2
ISSN 2632-0517

For EU product safety concerns:
IN TECH d.o.o., Prolaz Marije Krucifikse Kozulić 3, 51000 Rijeka, Croatia, info@intechopen.com or visit our website at intechopen.com.

IntechOpen Book Series
Veterinary Medicine and Science
Volume 7

Catrin Rutland is an Associate Professor of Anatomy and Developmental Genetics at the University of Nottingham, UK. She obtained a BSc from the University of Derby, England, a master's degree from Technische Universität München, Germany, and a Ph.D. from the University of Nottingham. She undertook a post-doctoral research fellowship in the School of Medicine before accepting tenure in Veterinary Medicine and Science. Dr. Rutland also obtained an MMedSci (Medical Education) and a Postgraduate Certificate in Higher Education (PGCHE). She is the author of more than sixty peer-reviewed journal articles, twelve books/book chapters, and more than 100 research abstracts in cardiovascular biology and oncology. She is a board member of the European Association of Veterinary Anatomists, Fellow of the Anatomical Society, and Senior Fellow of the Higher Education Academy. Dr. Rutland has also written popular science books for the public. https://orcid.org/0000-0002-2009-4898. www.nottingham.ac.uk/vet/people/catrin.rutland

Editor of Volume 7: Catrin Rutland
School of Veterinary Medicine and Science,
University of Nottingham, Nottingham, United Kingdom

Book Series Editor: Rita Payan-Carreira
University of Évora, Portugal

Scope of the Series

Paralleling similar advances in the medical field, astounding advances occurred in the Veterinary Medicine and Science in recent decades, fostering a better support to animal health and more humane animal production, a better understanding of the physiology of endangered species, to improve the assisted reproductive technologies or the pathogenesis of certain diseases, where animals can be used as models for human diseases (like cancer, degenerative diseases or fertility), and even as a guarantee of public health. Bridging the Human, Animal and Environmental health, the holistic and integrative "One Health" concept intimately associates the developments within those fields, projecting its advancements into practice.

This book series aims to tackle a variety of fields in the animal-related medicine and sciences, providing thematic volumes, high quality and significance in the field, directed to researchers and postgraduates. It aims to give us a glimpse into the new accomplishments in the Veterinary Medicine and Science field. By addressing hot topics in veterinary sciences, we aim to gather authoritative texts within each issue of this series, providing in-depth overviews and analysis for graduates, academics and practitioners and foreseeing a deeper understanding of the subject. Forthcoming texts, written and edited by experienced researchers from both industry and academia, will also discuss scientific challenges faced today in Veterinary Medicine and Science. In brief, we hope that books in this series will provide accessible references for those interested or working in this field and encourage learning in a range of different topics.

Contents

Preface XI

Section 1
Cardiovascular Disease and Genetics 1

Chapter 1 3
Canine Genetics and Genomics
by Edo D'Agaro, Andrea Favaro and Davide Rosa

Chapter 2 21
Diagnosis, Prognosis, Management, Treatment, Research and Advances in Canine Dilated Cardiomyopathy
by Siobhan Simpson, Kara-Zoë Kordtomeikel, Serena Wong, Samantha Bennison, Samir A.A. El-Gendy, Malcolm Cobb and Catrin Sian Rutland

Section 2
One Health 43

Chapter 3 45
The State of Knowledge on Intestinal Helminths in Free-Roaming Dogs in Southern South America
by Luciano Ritossa, Gustavo Viozzi and Verónica Flores

Chapter 4 79
Incrimination of Dog Vector of Cystic Echinococcosis and Impact of the Appropriate Dogs' Treatment
by Fatima Ezzahra Amarir, Abdelmohcine Aimrane and Abdelkbir Rhalem

Chapter 5 91
Importance of Yeasts in Oral Canine Mucosa
by Claudete Rodrigues Paula, Bianca Silva Navarro, Mário Mendes Bonci and Diana Costa Nascimento

Section 3
Oncology 111

Chapter 6 113
Small Animals Gut Microbiome and Its Relationship with Cancer
by Tatiane Moreno Ferrarias Epiphanio and Andreia A.F. Santos

Chapter 7 **141**
Canine Detection of the Volatile Organic Compounds Related to Cervical Cancer Cells
by Miriam Rodríguez-Esquivel, José Alejandro Sánchez-Pérez, Héctor Guerrero-Flores, Emmanuel Salcedo, Teresa Apresa-García, Ricardo López-Romero, Alejandra Valdivia-Flores, Florinda Jiménez-Vega, Rodrigo Aldariz and Mauricio Salcedo

Chapter 8 **155**
Canine Hepatic Carcinoma: Diagnoses and Treatments Via Global State-of-the-Art Approach and Traditional Chinese Veterinary Medicine
by Chih-Feng Chen, Shao-Wen Chiu, Lung-Kwang Pan and Chao-Hsun Chuang

Preface

Canine Genetics, Health and Medicine combines clinical and basic scientific research to present work being undertaken in fields including cardiovascular health, genetics, infectious and parasitic studies, and oncology. With nearly 1 billion dogs co-existing with us in the world, some living as companion animals and others free-roaming, understanding their needs and health is essential. Technology, research, and the availability and development of resources are expanding both what we know and what we understand about our canine friends. Therefore, we can push boundaries in clinical diagnosis and surgical methods, develop more pharmaceutical and surgical interventions, and better understand anatomy, physiology, pathology, and genetics. This enhanced information and innovation, alongside many other factors, enhance our diagnosis, prognosis, and treatment options and also expands our preventative medicine and general health and care opportunities.

The first section of this book is titled 'Cardiovascular Disease and Genetics.' Chapter 1 'Canine Genetics and Genomics' gives a general overview of how and why genetics is used in canine health, the advances being made, techniques being employed, and where the future lies for this field. It also explores areas in which genetics has an impact, ranging from hair color and breeding characteristics to heart disease and cancer. With increasing numbers of genetic tests becoming available commercially, both owners and veterinary professionals are often interested in not just genetic disorders but also in the vast array of information the canine genome holds such as breeding lines and phenotypic attributes and even delving into ancestry. This chapter introduces a wide range of uses for genetics, explains the background behind many of the techniques presently used, and looks at future directions of canine genomics. Chapter 2 'Diagnosis, Prognosis, Management, Treatment, Research and Advances in Canine Dilated Cardiomyopathy' shows an example of how genetic investigations have benefitted cardiovascular research. It highlights the latest advances in canine cardiomyopathy from pharmaceutical treatments to surgical procedures and healthy living advice. The chapter gives an overview of canine cardiomyopathy and discusses all the different techniques and standards presently used for diagnosis and prognosis, and the different treatment and management options available. It then continues with an in-depth look at the genetics of cardiomyopathy and its associations with other conditions. Finally, the chapter compares the genetic knowledge related to human cardiomyopathy with the information presently discovered in dogs and looks towards the future of both cardiomyopathy and advancements in technology in this area.

The second section, 'One Health,' concentrates on parasitic and yeast infections in canines Chapter 3 'The State of Knowledge on Intestinal Helminths in Free-Roaming Dogs in Southern South America', looks at the latest research undertaken on helminths. It covers issues such as the differences between urban and rural dogs, and how zoonotic diseases pose risks beyond the species of interest, and therefore also addresses some One Health issues. This chapter also covers important economic issues faced by people worldwide in relation to the effects of parasitic infections on canine health and medicine. Chapter 4 'Incrimination of Dog Vector of Cystic Echinococcosis and Impact of the Appropriate Dogs' Treatment', also investigates parasitic diseases from a One Health approach, concentrating on the latest knowledge on hydatidosis. This disease is not only an issue for dogs but is also a serious public health problem that can cause human morbidity and mortality as well as socio-economic impacts such as economic burden on the global livestock industry, therefore also affecting food security. Chapter 5 'Importance of Yeasts in Oral Canine Mucosa' covers areas such as the ecology and sources of yeast infection, how extrinsic and intrinsic changes in microorganisms can have serious implications for other diseases and disorders, and why oral health is important. Veterinary professionals are constantly seeking new ways to improve overall canine health and reduce disease and dentistry care provides an ideal opportunity to accomplish this. The chapter also goes into detail about several microorganisms, testing and identification methods, and treatment options.

The third section 'Oncology' begins with Chapter 6 'Small Animals Gut Microbiome and Its Relationship with Cancer,' which discusses the canine gut microbiome and its relationship with cancer. The chapter also explores the role of dogs as a model for studying humans. With 4 million new cancer cases a year, and an incidence rate similar to that in humans, oncology is an essential area of research. Chapter 7 'Canine Detection of the Volatile Organic Compounds Related to Cervical Cancer Cells' asks not what you can do for your dog but what your dog can do for you. Cervical cancer is a fundamental cause of cancer morbidity and mortality worldwide with rates in humans greater than 20 per 100,000. The World Health Organization (WHO) has highlighted that HPV vaccination, screening programs, and appropriate treatment services are required, and thus the focus of this chapter is on screening. It investigates the possibility of using dogs to recognise different volatile biomarkers emitted by cancer cells as a new diagnostic test that is low-risk, rapid, and non-invasive. Dogs have long been involved in diagnostic techniques, in addition to a vast number of other working roles, and this chapter explores their use not only in cervical cancer, but also in long-established diseases through to their latest employment in COVID-19 detection. Chapter 8 'Canine Hepatic Carcinoma: Diagnoses and Treatments Via Global State-of-the-Art Approach and Traditional Chinese Veterinary Medicine' looks at more modern medicine in addition to alternative medicines. The challenge with many alternative medicines is the lack of clinical trials and scientific testing. This chapter acknowledges these issues and brings together some of the research to date around hepatic carcinoma treatment in dogs. The chapter outlines the present diagnostic and treatment methods used for hepatic carcinoma and touches on the use of Chinese medicines, their use today, and their potential for tomorrow.

The common themes running throughout this book are gathering knowledge to help understand canine medicine and health, the subsequent impacts on human health and disease, and advances in technology and knowledge. Medicine has always been a constantly evolving field, yet we are also always working within differing socioeconomic backgrounds throughout the world. Research is constantly required to enhance diagnosis, prognosis, treatment, and clinical outcomes, to protect public health, economics, and food security, and most importantly to enhance animal welfare.

The editor would like to thank the authors, who are from several countries worldwide, for writing their chapters and sharing their expertise. She would also like to thank the team at IntechOpen, especially Karmen Daleta, Dolores Kuzelj, Mateo Pulko, Lucija Tomicic-Dromgool, Andrea Koric, and everyone else who works so hard to create these open-access books and resources.

Catrin Rutland
School of Veterinary Medicine and Science,
University of Nottingham,
Nottingham, United Kingdom

Section 1

Cardiovascular Disease and Genetics

Chapter 1

Canine Genetics and Genomics

Edo D'Agaro, Andrea Favaro and Davide Rosa

Abstract

In the past fifteen years, tremendous progress has been made in dog genomics. Several genetic aspects of cancer, heart disease, hip dysplasia, vision and hearing problems in dogs have been investigated and studied in detail. Genome-wide associative studies have made it possible to identify several genes associated with diseases, morphological and behavioral traits. The dog genome contains an extraordinary amount of genetic variability that distinguishes the different dog breeds. As a consequence of the selective programs, applied using stringent breed standards, each dog breed represents, today, a population isolated from the others. The availability of modern next generation sequencing (NGS) techniques and the identification of millions of single functional mutations (SNPs) has enabled us to obtain new and unknown detailed genomic data of the different breeds.

Keywords: canine genetics, canine genomics, bioinformatics, breeding, genetic diseases

1. Introduction

In recent years, genetic studies on dog genomics have multiplied worldwide. Currently, there are over 50 international laboratories which are involved in canine genome projects and several applications will be available in the near future from these studies. These new findings will improve our understanding of the selection process of the dogs and provide useful information for the study and control of genetic diseases.

2. Basic genetics

The single-control characters are influenced by genes located in a locus on one of the pairs of the chromosomes (78 in the dog) and have a binomial distribution. For example, the hair length in dogs is coded by two genes present at an autosomal locus. Short-haired animals have genotype LL (dominant homozygotes), while, long-haired animals have genotype ll (recessive homozygotes). From their mating originates short-haired animals with genotype Ll (heterozygotes), indistinguishable from short-haired parents. Even those characters that express different degrees of dominance, different from the Mendelian inheritance, are considered simple characters (e.g. incomplete or partial dominance). The simple characters are not influenced by the environment and, therefore, to each genotype corresponds a certain phenotype (P = G, where P = phenotype and G = genotype). The study of simple characters includes also multiple alleles (several alleles present in a population), pleitropy, association or linkage and incomplete penetrance. For characters with simple inheritance, it is easier to make selection than for multiple control characters. The multiple control characters are also called quantitative or polygenic characters. These characters are influenced by many genes distributed on several loci and they are influenced by environmental factors. The strong artificial selection exercised by man during the domestication process and during the creation of the different breeds has led to the setting of several characters. Color inheritance illustrates the case of separate loci that control the expression of the phenotype. The coat of dogs consists of two parts: top coat (protective function) and undercoat (heat-insulating function). Some breeds have no undercoat (e.g. Yorkshire). The color of the coat depends on the characteristics of the pigments contained in the medullary and cortical layers of the hair [1]. According to Willis [1], it is possible to explain all the colors by means of two chemical pigments: hemoglobin and melanin. More specifically, melanin is differentiated into eumelanin (black-brown) and pheomelanin (yellow-reddish). The synthesis of pigments in the hair of mammals depends on the interaction between the Agouti protein and the Melanocortin 1 receptor [2]. The coat colors in the dog are linked to the presence/absence of two types of melanin and their possible combinations. It is important to underline that melanin do not show a precise time of formation and they develop during the different phases of the fetal development and after birth [1]. The knowledge of the genetic inheritance of the morphological traits is very important in order to establish suitable selection objectives in the different breeds.

3. Relationship and inbreeding

Measurement of F coefficient (consanguinity) in a population can be considered as a measure of the increase in the proportion of homozygous individuals following an inbreeding mating (between relatives) [3]. The coefficient of consanguinity F can be calculated with the following methods: 1) pedigree 2) run of homozigosity (ROH); 3) genomic kinship matrix; 4) SNP genotyping [4, 5]. Inbreeding can occur in small closed populations due to mating between related animals. In a closed population, the decrease in the fraction of heterozygotes from one generation to the next may be referred to as ΔF. This value varies in relation to the size of the population: $\Delta F = 1/2Ne$ where Ne is the effective number or effective size of the population. In a population, Ne depends on the number of males (Nm), and on the number of females (Nf), in the following relationship:

$$1/Ne = 1/4\mathrm{Nf} + 1/4\,Nm; \Delta F = 1/8\mathrm{Nf} + 1/8\,Nm \tag{1}$$

The inbreeding coefficient, at a given t generation, can be calculated as a function of ΔF and t as:

$$\mathrm{F_t} = 1 - (1 - \Delta F)^{t} \tag{2}$$

which shows the decrease (ΔF) of heterozygotes that occurs at each generation following inbreeding [6]. Lewis *et al.* [7] reported for 221 breeds of the UK Kennel Club a Ne that varies between 23.8 of the Manchester terrier breed to 918 of the Borzoi breed and an average value of F equal to 0.06. The deleterious effects of inbreeding are universally known. They can be summarized briefly in the increase in the frequency of all genetic defects and abnormalities (reproductive sphere, resistance to diseases, longevity, etc.). These findings are based on the results of experiments carried out on different breeds and for several generations. Leroy *et al.* [8] showed that the increase in inbreeding in the population has an effect on individual survival and litter size of different breeds. Deleterious effects begin to occur when the value of F is about 0.375. Lower values are not to be considered dangerous. It is worth noting that this is the level of inbreeding that is achieved in only two generations of full sibling mating. For this reason, it is recommended to avoid mating between close relatives. Consanguinity is influenced by the number of individuals used per generation [9]. As a general rule, individuals whose numbers are lower in the breeding population they exert a proportionately greater effect on consanguinity. This is true both in relation to the male/female ratio (depend more on the number of males) and the different numbers of breeders in the various generations. The actual number of breeding animals is the parameter used in small populations to determine the expected inbreeding coefficient. Since the less numerous sex is the most important, the actual number of the population can be calculated even if the number of the larger sex is not known (e.g. 2 males and the number of females is assumed to be infinite: $1/Ne = 1/ 4Nf = 1/4\ (2) = 1/8 * F = 1/16 = 0.0625$). The family size is the number of offspring in each family who become parents in the next generation. In ideal conditions, the size of the population will remain constant in subsequent generations if each parent is replaced by another individual. In this case, the average number of offspring per parent is equal to 1 with an average family size of 2 (two parents). The Ne is also function of the variance of the family size. If males mate with more than one female, the number of offspring and thus the variance of the family size will differ between the two sexes. Several measures can be implemented to keep consanguinity within acceptable limits in the population: increase the number of breeders; mating of one male with a female (since the number within the sexes is the same, Ne will be maximized), reduce the variance size of the family (for a constant number of offspring for each family, the variance is equal to 0 and the Ne is double); avoid mating between siblings or cousins; avoid mating individuals in generations that overlap as inbreeding increases. If the management program includes the genetic improvement of one or more characters, selection must be carried out using selection indices that take into account of the level of relationship. The goal is to find the optimal number of offspring for each breeding animal and determine if a young animal (a candidate for selection) should be selected for breeding or not. This is done in an optimal way using the software EVA [10] that guarantees the achievement of the genetic progress and the maintenance an optimal genetic diversity in the population.

4. Breeding programs and strategies

The general actions to be taken in a program for the genetic improvement within a breed should include: 1) genomic identification and characterization of individuals, highlighting their potential in terms of their contribution to maintaining biodiversity, aptitude and use 2) monitoring of demographic parameters and assessment of the risk of reduced genetic variability 3) characterization and evaluation of the intra-breed genetic variability for proper management activities. Modern molecular techniques can be helpful for the improvement of management strategies, even for small breeds and for qualitative traits. The current hypothesis is to add molecular data to classical schemes (assisted selection) to improve their accuracy. The first step in planning an improvement program consists of: 1) a clear definition of the objectives 2) identification of the traits to be recorded 3) evaluation of the gene effect of the characters to be selected 4) estimate of the effect of the environment

Breed	DNA test	Physical test
Basenji	Fanconi	Eye assessment
		Hip score
	Progressive Retinal Atrophy	Thyroid
		Heart assessment
	Hemolytic anaemia	
	Pyruvate kinase deficiency	
	DNA inbreeding coefficient Factor	
	DNA identification Thyroid	
Border Collie	Neuronal Ceroid Lipofuscinosis	Elbow score
	Trapped Neutrophil Syndrome	Hip score
	Collie Eye Anomaly	Eye assessment
	Multi-Drug Resistance Gene 1	General vet check
	Imerslund-Grasbeck Syndrome	Chiropractor vet check
	Degenerative Myelopathy	Collie collaps
	Parentage (Orivet)	Hearing test
	Glaucoma	
German Shepherd	Degenerative Myelopathy	Hip score
	Ivermectin Sensitivity	Elbow score
	Long stock coat gene	
	Canine Renal Dysplasia	
	Dwarfism	
	Haemophilia	
Golden Retriever		
	Ichthyosis	Hip score
	Progressive Retinal Atrophy 1	Eye assessment
	Progressive Retinal Atrophy 2	Heart assessment
	Progressive Rod Cone Degeneration	Elbow score
		Dentition assesment

Table 1.
Genetic and physical testing used in genetic programs of common dog breeds.

(epigenetic effect) on the characters to be selected. In **Table 1** are reported the genetic and physical testing used in genetic programs of several dog breeds [11].

5. Genetic diseases and molecular diagnosis

In general, genetic diseases result from a mutation in a gene. In most cases, the mutations are traits that follow a simple Mendelian inheritance model (autosomal recessive, autosomal dominant or sex chromosome-linked character). Other hereditary diseases can be more complex and show reduced penetrance or multiple loci (multigenic disease). Genetic disorders can result from new mutations, but in

DOI: http://dx.doi.org/10.5772/intechopen.95781

most cases they result from old mutations passed on from one generation to the next. Mutated alleles can persist within a population for many reasons: 1. they can confer particular advantages in the state of heterozygotes; 2. the symptomatological signs can appear late 3. the mutation can be a recessive trait and therefore the defective allele can be spread in the population by healthy carriers. Without a mutation screening program, the carrier status can become evident only after the production of sick offspring.

The canine genome contains approximately 19,000 genes spread over 39 pairs of chromosomes (38 homologous chromosomes and 2 sex chromosomes). To date, nearly 400 hereditary diseases have been recognized in dogs. However, the precise ways in which these diseases are inherited are known for only about a third of them. In most cases, they are linked to autosomal recessive mutations. Bellumori *et al.* [12] report the prevalence of major genetic diseases in the United States for pure and mixed breeds. Pure breeds show more markedly some diseases including elbow dysplasia, cardiomyopathy, hypothyroidism and cataracts. The identification of the carriers can be implemented with the aid of two types of information: by pedigree or from a progeny test. In the first case, an animal showing the dominant phenotype (dominant phenotype) is known to be a carrier if one of the parents has the homozygous recessive genotype. In the second case, the farmer uses the information obtained from the offspring for the determination of the animal's genotype. Let us admit that a male is believed to be carrying a recessive allele. Special methods are required for the identification (and rejection) of carriers of the gene (suspected). This requires a reproduction test (test cross or progeny test) to determine whether the individual is dominant (suspected) or heterozygous. The genetic study of a hereditary diseases can follow additional strategies. Several genetic tests are now available for the identification of some hereditary disease [13]. The DNA-based diagnostic technique can be used to uniquely distinguish between sick and healthy subjects. These techniques allow the exclusion from reproduction of the carriers of frequent hereditary pathologies and they are a useful tool in validating the genealogical data reported in the pedigree.

6. Genomic analysis

6.1 Approach using candidate genes

The candidate gene approach consists in selecting a particular gene considered as the most likely site of a mutation. The main criteria for selecting a gene as a candidate are the following: 1) genes are selected because they are defective in similar animal species (usually humans or mice) 2) genes are selected based on their function. The analysis of the candidate gene consists in sequencing the entire gene and comparing two groups (healthy *vs* sick animals). However, the presence of a mutation in a gene is not in itself sufficient to identify the cause of the disorder. Unfortunately, for many genetic diseases the relative candidate gene has not been identified and very similar hereditary diseases can result from mutations on completely different genes. As an example, in the Bedlington terrier dog breed, the hereditary copper toxicosis is phenotypically identical to the Wilson's disease in humans. However, the gene involved in the human disease is not responsible for the disease in dogs. In conclusion, the approach with candidate genes has the advantage of allowing the identification of the specific mutation and therefore the creation of a targeted genetic tests.

6.2 Linkage analysis

The method of linkage analysis is based on completely different assumptions from the candidate gene approach. The main difference is that no assumptions are made about which gene is responsible for the disease, nor, more generally, the chromosomal tract involved. In this method, the whole genome is potentially subjected to analysis, without directing attention to any particular region. The search for the causal mutation takes place through the use of genetic markers whose chromosomal position is known. The more such markers are physically close to the mutation site, the more likely they will be co-inherited together with the mutation from one parental generation to the next. In a very simplified way, linkage analysis evaluates whether any of the variants of the markers appear in the population is associated with the presence of the disease. The ideal markers, and normally used to perform this type of study, are microsatellites, considered as practically ideal genetic markers because they are abundantly scattered throughout the genome and generally highly polymorphic. The number of microsatellites used to perform a linkage analysis is not fixed but generally the higher it is, the higher the probability that the study has success. This assumption derives from the fact that not directing attention towards specific genes and particular chromosomal portion, genome screening it must be as large as possible, i.e. it must contain the highest possible number of markers in order to understand the whole genome (so-called genome-wide screening). Generally, to perform a linkage study within a family tree informative are employed between 200 and 300 microsatellites using pedigrees with at least a hundred animals. For a given area of the genome, the probability of a recombination event occurring between a marker and a disease gene is directly proportional to their distance. The probability of occurrence of this event is expressed as a recombination fraction (θ). If θ is equal to 0.5, the marker and the disease gene are not linked and are therefore independently segregated. In other words, the probability that the marker and gene are inherited, associated or separated is identical. Conversely, if the marker and disease gene are linked together, the θ is less than 0.5. The lod score (Z) is the parameter which is used to estimate the linkage between 2 genetic loci. Z is the logarithm of the ratio between the probability that the 2 loci are linked ($\theta < 0.5$) and the probability that the 2 loci are randomly recombined ($\theta = 0.5$). Traditionally the linkage is accepted if the lod score is at least 3. Linkage analysis leads to the identification of a chromosomal region where the locus of the disease is probably located. The analysis must continue with the so-called refinement, that is, a further linkage analysis. Only later, the analysis proceeds through a gene candidate approach. All the genes of the region are identified and a sequence analysis is performed.

6.3 Genomic markers

6.3.1 Mitochondrial markers

Animal mtDNA is a cycular molecule ranging from 14,000 to 26,000 bp. The mtDNA codes for 13 proteins. Mitochondria contain most of the genes that code for cell energy production and electron transfer (NADH deydrogenase subunits, cytochrome oxidase subunits, ATPase 6 and 8, cytochrome b, rRNAr, RNA, 12S and 16S) [14, 15]. The choice of the sequence to be used for the genetic analysis depends on the phylogenetic hypothesis to be tested: D loop, sequences that evolve rapidly; cytocrome b, sequences that evolve moderately; Cytocrome oxidase I, sequences that evolve slowly. The mitochondrial control region (CR) sequence is the most popular marker. The mtDNA is uniparental (maternal line), characterized by a high evolution rate (5–10 times higher than nuclear genes) and the lack of introns and recombinations. The mtDNA is used to clarify the direction of hybridization and the incidence of introgression. In the case of hybridization, erroneous inferences can be obtained only using the evolutionary history of the females. In phylogeographic studies, information from various loci of the nuclear genome are also included [16–18]. The use of both parents allows a better analysis of the population structure.

6.3.2 Microsatellite markers

Nuclear microsatellites (one to six in tandem repeated nucleotides) are used in population genetics for the description of the population structure and kinship identification [19]. The reason for the wide use of microsatellites is due to the fact that are co-dominant, multi-allelic, highly reproducible and with a high resolution. The information per locus is about 10 times more informative than SNP markers. The most common repeats are di, tri and tetra-nucleotides. Microsatellite loci with a di-nucleotide motif are generally used, since they are easier to isolate and high density (on average every 30–50 kb) [20]. Microsatellites are also known as SSR (Simple Sequence Repeats) or STRs (Short Tandem Repeats). The maximum length is about 200 bp. Microsatellites are distributed throughout the genome with greater prevalence in non-coding regions. They are neutral in terms of selection. The typical problems encountered in the genotyping analysis are: homoplasy (condition of equality in the type and number of microsatellite repeats between two alleles) [21]; stutters (in the form of allelic pre-peaks); null alleles (NA) (possible mutations in the pairing site of the primers can prevent the pairing to the target sequence, causing the non-amplification of some alleles. The genetic analysis of microsatellites produce the following data: the distribution of allele frequencies for each microsatellite locus, the percentage of expected (H_E) and observed (H_O) heterozygosity, the estimates of the F_{st} values; Nei distances; conformity to the Hardy–Weinberg equilibrium (HWE) of the allele frequencies for each locus.

6.4 Next generation sequencing (NGS)

Starting in the 2000s, the analysis of SNPs led to the beginning of a new era in molecular genetics. The direct study of the genome using SNPs markers allows to integrate the genealogical information and to obtain high levels of accuracy in the estimation of the main genetic parameters of the population. The development of new sequencing techniques has made it possible to study the consequences of gene flow using a larger number of markers. At the beginning, the Sanger's technology was used to sequence the genomes of different animal species. This sequencing technique produces reads (>700 bp) with a very low error (<0.01%) and high cost (>600 US $ per Gb). This technique was subsequently improved through the use of the Celera assembler with a significant reduction in time and costs. New generation sequencing technologies (Next Generation Sequencing - NGS), also known as High Throughput Sequencing (HTS) technologies, have evolved rapidly offering an ever greater number of sequenced bases at a lower cost. In 2006, the first second-generation NGS technologies (Second-Generation Sequencing - SGS) appeared. Illumina (MiSeq, HiSeq and NovaSeq) is the most popular platform, due to its high performance and low cost. This technology is based on the fragmentation of DNA, amplification in multiple reactions in parallel, obtaining short reads, between 100 and 300 bp. Depending on the library, it is possible to sequence only one end of the fragment, single reads (single end) or both ends. The distance between the read pairs is called insert size (mate pair (2–5 kb); paired end (<1 kb)). Since 2013, the third-generation NGS techniques emerged, also known as the Single Molecule Sequencing (SMS) method. Single molecule sequencing produces long reads with higher costs (>2000 US$ per Gb). These techniques do not require the library amplification step and they are capable of directly sequencing a single DNA molecule, without applying any enzymatic or hybridization process. The main platforms of the third generation are Pacific Biosciences (PacBio) and Oxford Nanopore Technologies (ONT). These platforms produce longer reads than the previous ones (5–50 kb) but have a much higher error rate. The Pacbio platform routinely generates reads with an N50 > 1 Mb and it has recently reduced the error rate with a new technique (circular consensus sequencing) and the production of high fidelity reads of 15 Kb. The most popular softwares used for the bio-informatic analysis are Canu; Marvel and Mecat Flye. Then, results obtained are cleaned with some software such as Racon; Nanopolish and Pilon. **Figure 1** shows an example of workflow using long reads.

After identifying the putative protein coding regions (CDSs), UCEs (Ultra Conserved Elements), it is possible to infer the correct reading pattern (Open Reading Frames, ORF) and translate the nucleotide sequences into amino acids [22]. In this way, we will obtain the set of predicted proteins encoded by the study genome. BLAST (nucleotide, protein, translated, genomes), HMMER or InterProScan databases can be used to functionally annotate these proteins. InterProScan provides the information on functional processes (GO terms) and metabolic pathways (KEGG). Once the functional and structural annotation has been obtained, the analysis of the functional elements of interest such as polymorphic positions or genes with differential expression can be performed. **Figure 2** shows an example of workflow for the genomic annotation analysis.

Orthmcm, Orthofinder; EggNog sofwares can be used for the homology analysis. Several studies, in recent years, have shown that the best way to understand complex systems (for example diseases) is to combine different omic data together. **Figure 3** shows a detailed analysis using omic data (genomic, transcrptomics, proteomics and metabolomics).

6.5 Reduced representation genome sequencing (RRGS)

Several new techniques have been developed in the last decade. The most popular is the restriction-site-associated DNA sequencing (RAD-Seq) [23] and the genotyping by sequencing method (GBS) [24]. The main advantage of RRGS methods is that it reduces the cost of analysis with an high coverage compared to the traditional sequencing methods. The *de novo* analysis does not require a priori knowledge of the reference genome sequence. Several applications of the RAD-Seq methods have been reported: population genetics studies (phylogenetic and

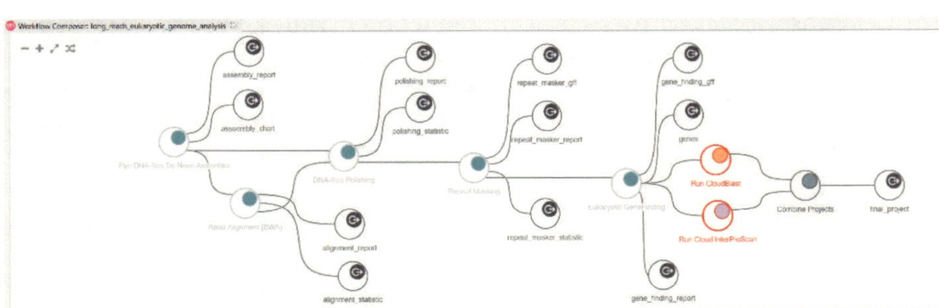

Figure 1.
Example of NGS bioinformatic analysis (long read sequencing).

DOI: http://dx.doi.org/10.5772/intechopen.95781

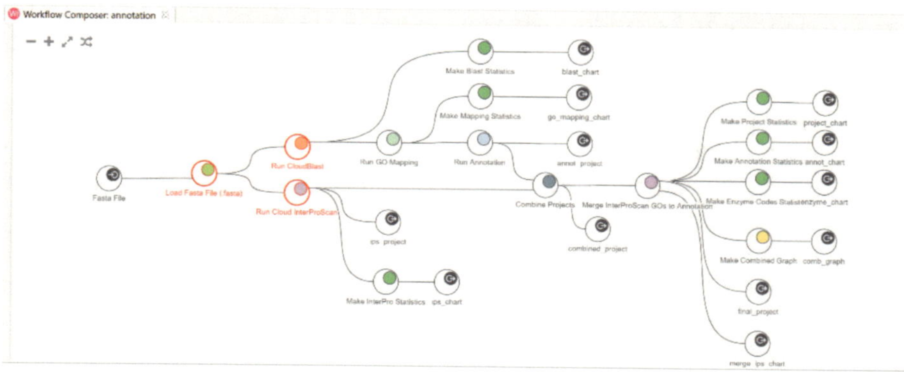

Figure 2.
Example of NGS annotation analysis.

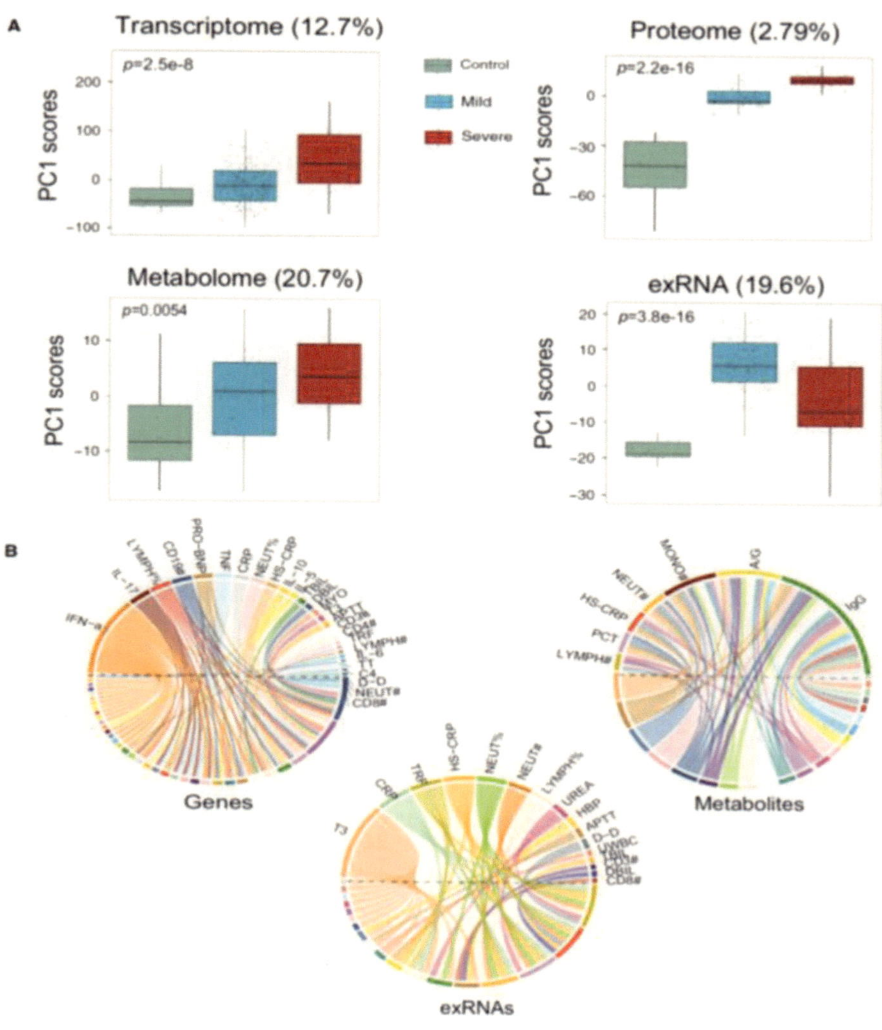

Figure 3.
Example of OMICs analysis (genomics,transcrptomics, proteomics and metabolomics).

phylogeographic), linkage mapping (fine scale) and genome scaffolding [25]. To avoid or reduce the bias, some variations of the original RAD Seq protocol have been proposed: ddRAD, ezRAD, 2b-RAD. Classic RAD reads are obtained between the restriction site and a random site while the ddRAD reads are obtained between two restriction sites. In particular, the ddRAD-SEq method increases the number of samples per sequencing line and develops a tagging approach by combining pairs of adapters. Another advantage is the selection of the fragment sizes. This reduces duplicate sampling of a region, thus requiring only half the reads to effectively achieve high levels of confidence for each SNP associated with a restriction enzyme cleavage site. All these properties make the ddRAD-Seq method robust, allowing to search for a smaller number of reads. The bio-informatic analysis of RAD-Seq data includes the following phases: quality control, trimming, reference genome or *de novo* mapping methods, SNP filtering/annotation. The results of RAD-Seq analysis are analyzed with different softwares such as Stacks, Ig-Tree, Uneak (Tassel), Pyrad; Ddocent; 2brad and Aftrrad. The most popular software is the Stacks program. RRL methods, in relation to the production of short reads, are not very useful for the construction of phylogenetic trees but are generally used for the analysis of SNPs.

6.6 Genome-wide genotyping arrays

In the recent years, the availability of massive genomic data obtained from the last generation sequencing techniques allowed the efficient identification of a large number of SNPs [26]. The GWAS is a method of investigation that allows to examine the entire genome by analyzing the single nucleotide polymorphism of genomic markers (SNPs) with the use of high density SNP arrays [27] (the last versions Illumina Canine HD SNP 170 K have hundreds of thousands of SNPs distributed throughout the genome). The study identified the genetic structure of the populations present in Italy and the selection signatures. Reduction of genotyping costs is achieved using inference methods such as the imputation. Imputation techniques allow to transfer information from DNA from high density bead chips to low density ones.

6.6.1 Genome-wide association studies

The genome-wide association studies (GWAS) have been proposed as an effective approach for the identification of many causative mutations and genetic factors that constitute the main traits. Unlike linkage studies, which consider the phenomenon of inheritance of chromosomal regions linked to the presence of a trait within a family, association studies consider instead the difference between the frequency of SNPs affecting the trait of interest. Association studies may be conducted through two approaches: direct and indirect. A direct association study is to catalog and test one by one all the possible causal mutations. However, the direct approach presents some practical problems. This strategy involves genome-wide identification of all genes (up to 19,000 genes) as well as of all SNPs. For these reasons, the use of the direct method is limited to a few cases and it has almost always replaced with the application of the indirect method. The indirect strategy avoids the need to catalog all mutations that could potentially give predisposition to a given trait and instead relies on the association between a giver phenotype and markers located near a strategic locus. These associations are obtained from studies of linkage disequilibrium (LD) between marker loci. The indirect strategy, then employs a dense map of polymorphic markers to explore the genome in a systematic way. The choice of markers differentiates further the indirect approach in two different strategies. In the first, markers are chosen very close to exon regions of known genes. The second employs markers located in large regions, virtually anywhere in the genome, thus considering the chromosomes in their entirety, including intronic regions. The choice of the marker falls on bi-allelic SNPs because of their high frequency in the animal genomes, for the low rate of mutation and for the ease with which it can be analyzed. Linkage means the presence of genes in closed loci on the same chromosome. LD is a combination of alleles at two or more loci that occurs more often than it does happen by chance. Two markers are in LD when they occur together in the same individual more frequently than would be expected by chance. The presence of a LD thus indicates co-segregation of two markers. In generally, the LD between two SNPs decreases with the physical distance and the extent of LD varies strongly among the regions of the genome. LD analysis is a valuable tool for fine mapping. Doherty [28] conducted an GWAS analysis using 9700 SNPs on 72,000 dogs (63 breeds). Eight SNPs were significantly correlated with the live weight and five SNPs with cancer mortality. Plassais [29] analyzed the genomes (WGS) of 722 dogs and used the Illumina canine HD SNP BeadArray to identify over 91 million SNPs. In this way the main SNPs coding for body weight and main morphological characters were identified. In **Table 2** is reported an example of SNP genotyping using a SNP chip array in dogs [30].

6.7 Scans for selective sweeps

The domestic dog is thought to be the most recent species of the canine family, within which three phylogenetic groups, or clades, are distinguished: the domestic dog belongs to the same clade as the gray wolf, coyote and jackals [31]. It is thought that the dog appeared about 40,000 years ago, and that the first steps in its domestication took place in East Asia [32]. Most of the domestic breeds we know today, however, are the result of human selection over the past two or three centuries. Many of the most popular modern breeds were created in Europe in the 19th century. Some of the breeds were already present in the ancient world as the greyhound and the dog of the pharaohs. Studies conducted at the genomic level have highlighted a stratification of genetic variability within dog breeds. The recent sequencing methods and the use of SNP arrays allow the screening of the whole genome for the presence of signatures of selection. Sequencing data are aligned to the reference genome to identify selective sweeps. The presence of genes with

Genomic analysis
Illumina CanineHD SNP chip (San Diego, CA)
Genotype SNP calls using Illumina's Genome Studio
Selection of samples with a >90% SNP call rate
SNPs with Gentrain scores >0.4
Minor allele frequency >1%
Bio-statistical analysis
FlashPCA - Principal components (PCs)
Admixture - two to ten adjusted cluster ancestry models
Beagle – calculation of IBD haplotype sharing analysis and phasing
VCFtools – calculation of the inbreeding coefficient
TreeMix – construction of a maximum likelihood tree - windows of 1000 SNPs using the flags -bootstrap and -k 1000 functions
R studio – construction of graphs and plots for all the analyses

Table 2.
SNP genotyping using a SNP chip array in dogs.

Dataset of 268 dogs representing 130 breeds
Phenotypes used in the study: canids catalog, kinship, aggressiveness, boldness, bulky, drop ears, furnishing, hairless, height, large ears, lengh of fur, life span, long legs, muscled, tail curl, weight, white chest, white head
GWAS
Samples with ≥10x coverage, selecting two males and two females
Gemma - linear-mixed model methods; elimination of variants with missing value >1
R Studio – Manhattan correlation and box-plots
Identification of positively selected genes
Vcftools60
Beagle - infer the haplotype phase
Xpclr - phased genotype input; non-overlapping windows (50 kb), 600 SNPs within each window; correlation level cutoff of 0.95.
XP-EHH - splitting the genome into non-overlapping segments of 50 kb

Table 3.
Example of GWAS and selective sweep analysis in dogs.

a large number of outliers indicates a positive or negative effect of selection. A genome scan approach can be used to distinguish genome-wide processes (expected to mainly reflect demographic histories) from processes at individual loci. Genome scans may suffer from inflated numbers of false positives under hierarchical spatial structure coupled with isolation by-distance dynamics. In the case of positive selection, there is an increase in the fitness of the population due to a new (or rare) mutation. In the case of hard sweeps, there is an increase in the frequency of some variants and in the linkage disequilibrium. Kim et al. [33] compared 127 dogs (sport-hunting vs. terrier) for sporting characteristics. Results of the study showed the main SNPs (cardio-circulatory, muscular and neuronal systems) and selection signature that are involved in the sport-hunting breeds. In **Table 3** is reported an example of GWAS and selective sweep analysis in dogs [29].

7. Genome applications in the canine sector

The canine genome project was launched in the early 1990s. After some preliminary results, in 2003, a fist sequence of the dog's genome was obtained from a female boxer which is now the reference sequence for the dog [34]. The availability of a high quality canine genome has revolutionized the way in which geneticists operate. The first version of the boxer's genome, carried out with a coverage of 7.5x, covered nearly 99 percent of the animal's genome. The genome sequence provided a first description of the organization, number of genes and the presence of repeated elements. To some surprise, they found a high presence of short interspersed nuclear elements (SINEs) throughout the genome, sometimes located in locations from which they could affect gene expression. For example, the insertion of a SINE into the gene encoding the hypocretin receptor (a neuropeptide hormone found in the hypothalamus) causes narcolepsy in the Doberman. Similarly, the insertion of a SINE element into the silv gene, which is known to be linked to the pigmentation process, is responsible for a particular mottled color called merle. The 2003 sequence comprises approximately 2.4 billion of bases and revealed the existence of approximately 19,000 genes. For about 75% of genes, the homology (resulting from shared ancestral material) between the dog, man and mouse is very high. The study also found that many genes have no gaps in their sequence, which is beneficial if you would like to study the correlation between a given gene and a disease. During its evolution, the dog's genome has accumulated more than two million of SNPs. These markers are proving crucial in understanding the role of genetic variability within one breed and in different breeds. SNPs, analyzed by means of DNA microarrays or bead arrays, can make an important contribution to GWAS (association studies) aimed at identifying the genes responsible for complex traits in dogs. A microarray with around 170,000 SNPs is currently available. By comparing data from dogs with a certain disease with healthy individuals, it is possible to quickly identify the genes responsible for the disease. Dog breeds differ not only in the overall body size but also in leg length, head shape and many other morphological characteristics. In the dog, the phenotypic variability of several traits is very high compared to the other living terrestrial mammals. The first molecular study on the genetic aspects of dog morphology was conducted at the University of Utah [35, 36]. Called Georgie Project (in memory of a dog), the study focused on the Portuguese water dog breed, ideal for this type of study because it comes from a small number of ancestors. In the project, DNA samples of more than a thousand dogs were collected. A completed genome scan using 500 microsatellite markers was carried out. For these animals, in addition to the genealogical and medical data, more than 90 anatomical measurements were obtained from a series of five radiographs taken on each animal during the first phase of the study. Based on the analysis of these data, four primary main components (CP) have been identified (**Figure 4**).

The analysis of the genome scans and principal components (CPs) revealed 44 putative QTLs (quantitative trait loci associated with a particular quantitative trait) on 22 chromosomes. QTLs are identified by means of a complicated statistical analysis and identify the genome regions that contribute to the expression of a certain trait. Of particular interest is the gene CFA15 on chromosome 15 which showed a strong association with the body size. Although, it is only one of seven loci thought to affect the body size, it was chosen as the starting point. To find the gene CFA15, several SNPs were identified and then the resulting set of genome-wide markers were genotyped. The distribution of these markers showed a single peak near the insulin-like growth factor-1 (IGF 1) gene, which codes for insulin-like growth factor which is known to code for the body size in humans and mice. IGF 1 was analyzed in detail, discovering that there are only two specific combinations of alleles (called haplotypes) and one of them is present in 96% of the population. The haplotype associated with the small size was called B, while the one associated

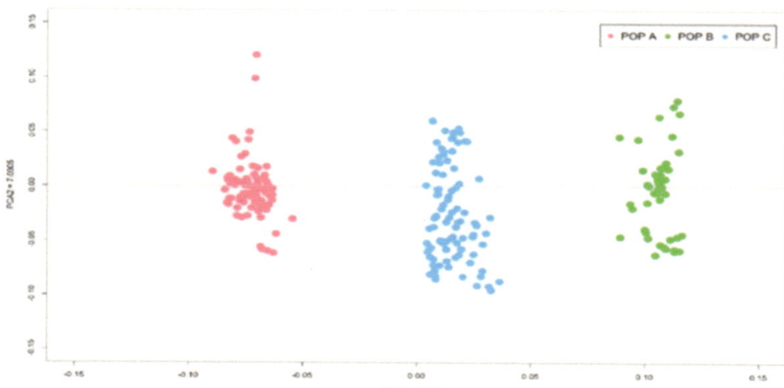

Figure 4.
Example of PCA (principal component analysis) of genotypic data (autosomal) of three dog populations.

with the largest size was called I. Homozygous dogs for the haplotype B showed a smaller average body size while, dogs homozygous for I were larger. Heterozygous dogs showed an intermediate size. The Georgie Project is important for the number of genes discovered. In addition to the genes related to the head shape, body size, leg length and many other traits, additional genes were discovered that control the sexual dimorphism [37, 38]. This dimorphism is observed in almost all mammals but its mechanisms it is not yet fully known. Indeed, it was found a gene on chromosome 15 which interacts with other genes to make males larger and females smaller. On average, females of the Portuguese water dog breed are 15% smaller than the males.

8. Future perspectives

In the past fifteen years, tremendous progress has been made in dog genomics [39–41]. Several genetic aspects of cancer, heart disease, hip dysplasia, vision and hearing problems in dogs have been investigated and studied in detail. Genome-wide associative studies have made possible to identify several genes associated with diseases, morphological and behavioral traits. The Dog10K project will produce 10,000 new dog genomes (20x) within five years [42]. The mapping of disease-associated genes will hopefully lead to the production of new genetic tests and improve the management of running breeding programs, which in turn will produce healthier and longer-living dogs. It will be easier to select for specific physical traits such as the size or coat color. Finally, perhaps we will be able to identity which genes are responsible for the typical behaviors of each breed.

Author details

Edo D'Agaro*, Andrea Favaro and Davide Rosa
Department of Agricultural, Food, Environment and Animal Sciences,
University of Udine, Udine

*Address all correspondence to: edo.dagaro@uniud.it

References

[1] Willis MB,. Genetics of the dog. Howell Book House. 1989; New York, USA.

[2] Berryere TG, Kerns JA, Barsh GS, Schmutz SM. Association of agouti allele with fawn or sable coat color in domestic dogs. Mamm Genom. 2005; 16: 262-272.

[3] Mellanby RJ, Ogden R, Clements DN, French AT, Gow AG, Powell R, Corcoran B, Schoeman JP, Summers KM. Population structure and genetic heterogeneity in popular dog breeds in the UK. Vet J. 2013; 196(1):92-97.

[4] Zhang Q, Calus MPL, Guldbrandtsen B, Lund MS, Sahana G. Estimation of inbreeding using pedigree, 50k SNP chip genotypes and full sequence data in three cattle breeds. BMC Genet. 2015; 16:88.

[5] Wang J. Pedigrees or markers: Which are better in estimating relatedness and inbreeding coefficient? Theor Popul Biol. 2016; 107: 4-13.

[6] Jansson M, Laikre L. Pedigree data indicate rapid inbreeding and loss of genetic diversity within populations of native, traditional dog breeds of conservation concern. Plos one. 2018; 13(9): e0202849.

[7] Lewis TW, AbhayaratneBM, BlottSC. Trends in genetic diversity for all Kennel Club registered pedigree dog breeds. Canine Genet Epidemiol. 2015; 2:13.

[8] Leroy G, Phocas F, Hedan B, Verrier E, Rognon X. Inbreeding impact on litter size and survival in selected canine breeds. Vet J. 2015; 203:74-78.

[9] Leroy G. Genetic diversity, inbreeding and breeding practices in dogs: Results from pedigree analyses. Vet J. 2011; 189:177-182.

[10] Berg P, Nielsen J, Sørensen MK. EVA: Realized and predicted optimal genetic contributions. WCGALP. 2006; 246.

[11] Czerwinski V, McArthur V, Smith B, Hynd P, Susan Hazel S. Selection of breeding stock among Australian purebred dog breeders, with particular emphasis on the dam. Animals. 2016; 6: 75.

[12] Bellumori TP, Famula TR, Bannasch DL, Belanger JM; Oberbauer AM, Bellumori TP, Famula TR Bannasch DL, Belanger JM, Oberbauer AM. J Am Vet Med Assoc. 2013; 242:1549-1555.

[13] Oberbauer AM, Belanger JM, Bellumori T, Bannasch DL, FamulaTR. Ten inherited disorders in purebred dogs by functional breed groupings. Canine Genet Epidemiol. 2015; 2:9.

[14] Avise JC. Perspective: conservation genetics enters the genomics era. Cons Genet. 2010; 11:665-669.

[15] Duleba A, Skonieczna K, Bogdanowicz W, Malyarchuk B, Grzybowski T. Complete mitochondrial genome database and standardized classification system for *Canis lupus familiaris*. Forensic Sci Int-Gen. 2015; 19:123-129.

[16] Ersmark E, Klütsch CFC, Chan YL, Sinding M-H S, Fain SR, Illarionova NA. From the past to the present: Wolf phylogeography and demographic history based on the mitochondrial control region. Front Ecol Evol. 2016; 4:134.

[17] Hendrickson M, Remm J, Pilot M, Godinho R, Stronen AV, Baltrūnaité L. Wolf population genetics in Europe: a systematic review, meta-analysis and suggestions for conservation and management. Biol Rev. 2017; 92:1601-1629.

[18] Toth G, Gaspari Z, Jurka J. Microsatellites in different eukaryotic genomes: survey and analysis. Genome Res. 2000; 10:967-981.

[19] Zane L, Bargelloni L, Patarnello T. Strategies for microsatellite isolation: a review. Mol Ecol. 2002; 11:1-16.

[20] Estoup A, Rousset F, Michalakis Y, Cornuet J-M, Adriamanga M, Guyomard R. Comparative analysis of microsatellite and allozyme markers: a case study investigating microgeographic differentiation in brown trout (*Salmo trutta*). Mol Ecol. 1998; 7:339-353.

[21] Grabherr M, Haas B, Yassour M,LevinJ, Thompson D, Amit I, Adiconis X, Fan L, Raychowdhury R, Zeng Q.Trinity:reconstructing full-length transcriptome without a genome from RNA-Seq data. Nat Biotechnol. 2011; 29(7):644-652.

[22] Martin SH, Davey JW, Jiggins CD, Evaluating the use of ABBA-BABA statistics to locate introgressed loci. Mol. Biol. Evol. 2015; 32:244-257.

[23] Lu F, Lipka AE, Glaubitz J, Elshire R, Cherney JH, Casler MD. Switchgrass Genomic Diversity, Ploidy, and Evolution: Novel Insights from a Network-Based SNP Discovery Protocol. Plos Genet. 2013; 9(1):e1003215.

[24] Baird NA, Etter PD, Atwood TS et al. (2008) Rapid SNP discovery and genetic mapping using sequenced RAD markers. Plos one. 2008; 3: e3376.

[25] Daetwyler HD, Hickey JM, Henshall JM, Dominik S, Gredler B. Accuracy of estimated genomic breeding values for wool and meat traits in a multi-breed sheep population. Anim Prod Sci. 2010; 50:1004-1010.

[26] Ramos AM, Crooijmans RPMA, Affara NA, Amaral A J, Archibald AL. ration sequencing technology. Plos one. 2009; 4: e6524.

[27] Meuwissen THE. Genomic selection: marker assisted selection on a genome wide scale. J Anim Breed Genet. 2007; 124: 321-322.

[28] Doherty A, Lopes, Ford CT, Monaco G, Guest P, de Magalhães JP. A scan for genes associated with cancer mortality and longevity in pedigree dog breeds. Mamm Genome. 2020; 31:215-227.

[29] Plassais J, Kim j, Davis BW, Karyadi DM, Hogan AN, HarrisAC, Decker B, Parker HG, Ostrander EA. Whole genome sequencing of canids reveals genomic regions under selection and variants influencing morphology. Nature Comm. 2019; 10:1489.

[30] Ali MB, Evans JM, Parker HG, Kim J, Pearce-Kelling S, Whitaker DT, et al. (2020) Genetic analysis of the modern Australian labradoodle dog breed reveals an excess of the poodle genome. PLoS Genet. 2020; 16(9): e1008956.

[31] Vonholdt BM, Pollinger JP, Lohmueller KE, et al. Genome-wide SNP and haplotype analyses reveal a rich history underlying dog domestication. Nature. 2010; 464:898-902.

[32] Wayne RK, Ostrander EA. Lessons learned from the dog genome. Trends in Genet. 2007; 23:11.

[33] Kim J, Williams FJ, Dreger DL, Plassais J, Davis BW, Parker HG, Ostrander EA. Genetic selection of athletic success in sport-hunting dogs. Pnas. 2018; 115:30.

[34] Parker HG, Dreger DL, Rimbault M, Davis BW, Mullen AB, Carpintero-Ramirez G, Ostrander EA. Genomic analyses reveal the influence of geographic origin migration, and hybridization on modern dog breed development. Cell Rep. 2017; 19: 697-708.

[35] Chase K, Carrier DR, Adler FR, Jarvik T, Ostrander EA, Lorentzen TD, Lark KG. Genetic basis for systems of skeletal quantitative traits: Principal component analysis of the canid skeleton. Pnas. 2002; 99(15):9930-9935.

[36] Chase K, Carrier DR, Adler FR, Ostrander EA, Lark KG. Interaction between the X chromosome and an autosome regulates size sexual dimorphism in Portuguese Water Dogs. Genome Research. 2005; 15:1820-1824.

[37] Lindblad-Toh K., Wade CM, Lander ES. Genome sequence, comparative analysis and haplotype structure of the domestic dog. Nature. 2005; 438:803-819.

[38] Rimbault M, Beale HC, Schoenebeck JJ, Hoopes BC, Allen JJ , Kilroy-Glynn P, Wayne RK, Sutter NB, Ostrander EA. Derived variants at six genes explain nearly half of size reduction in dog breeds. Genome Res. 2013; 23:1985-1995.

[39] Parker HG, Meurs KM, Ostrander EA. Finding cardiovascular disease genes in the dog. J Vet Cardiol. 2006; 8:115-127.

[40] Parker HG and Ostrander FA. Canine genomics and genetics: Running with the pack. Plos Genet. 2005;1(5):e58, 2005.

[41] Dreger DL, Davis BW, Cocco R, Sechi S, Di Cerbo A, Parker HG, et al. Commonalities in development of pure breeds and population isolates revealed in the genome of the Sardinian Fonni's Dog. Genetics. 2016; 204:737-55.

[42] Ostrander EA, Wayne RK, Freedman A H, Davis BW. Demographic history, selection and functional diversity of the canine genome. Nat Rev Genet. 2017; 18:705-720.

Chapter 2

Diagnosis, Prognosis, Management, Treatment, Research and Advances in Canine Dilated Cardiomyopathy

Siobhan Simpson, Kara-Zoë Kordtomeikel, Serena Wong, Samantha Bennison, Samir A.A. El-Gendy, Malcolm Cobb and Catrin Sian Rutland

Abstract

Dilated cardiomyopathy involves enlargement of the ventricular chamber and systolic dysfunction. The reduction in quality of life and increased levels of congestive heart failure, combined with the high diagnosis rate within the canine population, highlights the need for research into this disorder. This chapter looks at prevention, diagnosis, prognosis, and treatment of dilated cardiomyopathy. It details the disease pathology and physiology through to present clinical practices and studies to support prevention and treatment. This chapter also looks at the research being undertaken to further understand cardiomyopathies in dogs and develop new interventions. This ranges from fatty acids profiles to genetics and even personalized medicine and comparisons with human cardiomyopathy.

Keywords: Dilated cardiomyopathy (DCM), Canine, Echocardiography, Holter Monitoring

1. Introduction

DCM is characterised by ventricular chamber enlargement and systolic dysfunction which often leads to congestive heart failure [1]. The aetiology of DCM is complex. Genetic factors, myocardial ischemia, hypertension, toxins, infections and metabolic defects have been implicated [2]. DCM is the most common specific heart disease diagnosis within the Swedish insured canine population, following the more general diagnosis of cardiomyopathy, accounting for 10% of the cardiac diagnoses [1]. Prevention of disease requires a thorough understanding of the underlying causes of disease. Where the causes of disease are understood it can allow for modifications in diet, behaviour, and/or preventative medicine to be prescribed, or risk-reducing surgery to be undertaken where appropriate [3–5]. The underlying causes of non-communicable diseases are varied, with a wide range of environmental and genetic factors contributing to disease [6–11]. In most cases a combination of interacting factors contribute to disease risk, initiation, and progression [12–15].

2. Diagnosis and prognosis of canine cardiomyopathy

Dilated cardiomyopathy (DCM) is a significant cause of congestive heart failure in dogs, characterised by the enlargement and impaired contraction of the left or both ventricles [1–3]. The development of DCM can be classified into three main stages [4, 5]. In Stage 1, the heart appears normal, with no clinical evidence of heart disease and often includes dogs that are genetically predisposed to DCM [6]. Stage 2 (the preclinical or occult phase) is characterised by morphological and electrical cardiac changes with a prolonged period without overt clinical symptoms. Stage 3 (the overt phase) includes clinical signs of congestive heart failure [1–3, 7]. It should also be noted that adult DCM clinical signs can vary between breeds.

The gold standard approach to DCM diagnosis relies on echocardiographic and 24-hour electrocardiographic (ECG) assessments (**Figure 1**), in conjunction with monitoring clinical presentation and signalment [7–9]. The most common early clinical signs include exercise intolerance and heart murmurs/irregular heart rhythms. As the condition develops pulmonary congestion edema may develop and abdominal fluid accumulation and/or pleural effusion may be present. Notable other signs including weakness, inappetence, weight loss, breathlessness, coughing, increased breathing rate, collapse and lethargy are more frequent in dogs with heart failure caused by DCM, as is sudden death [3]. Congenital or acquired cardiac diseases with similar presentations to DCM must be also be excluded [10, 11].

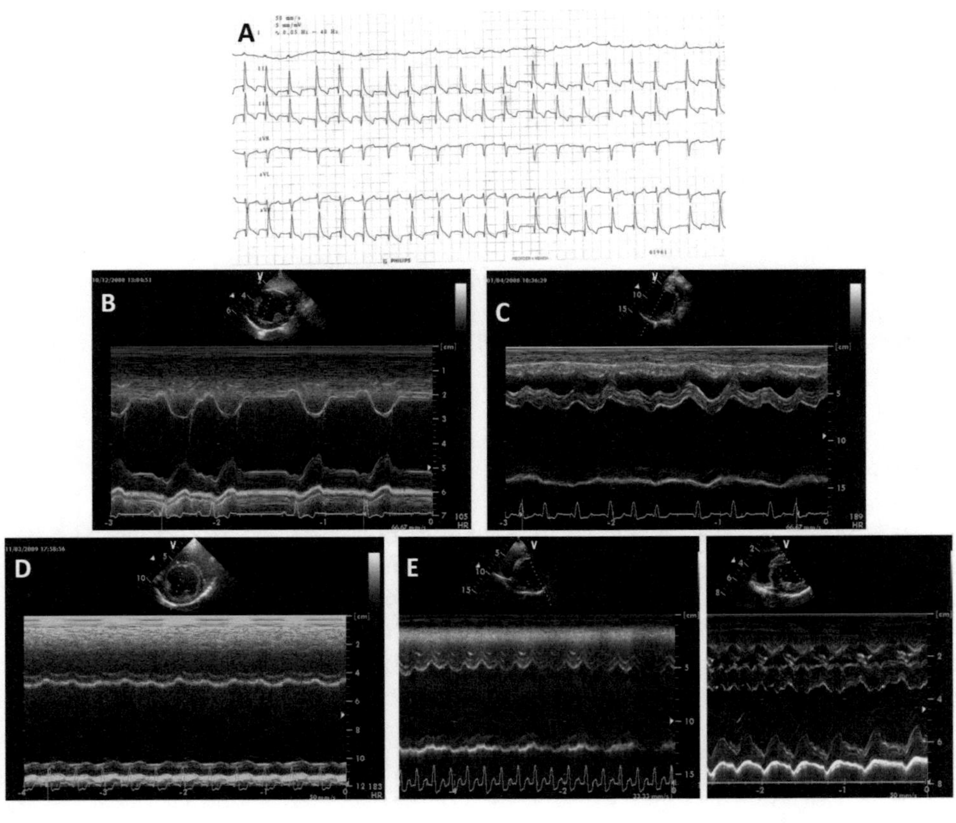

Figure 1.
A) Six lead ECG showing fast atrial fibrillation in a dog with dilated cardiomyopathy (DCM). Motion-mode echocardiogram from dogs that were classified as B) clinically normal, C) DCM showing dilation of the left ventricle and irregular filling associated with atrial fibrillation, D) DCM showing dilation of the left ventricle and a sinus tachycardia, and E) DCM (Left) alongside a motion-mode echocardiogram from a normal dog (Right), note the difference in the size of the ventricles and the amount of movement in the walls.

Echocardiography is used to assess left ventricular (LV) dimensions and function, where a dilated ventricle, based on M-mode or 2D measurements, with reduced contractility is indicative of DCM [9, 12]. M-mode is a time motion technique displaying the movement of structures over several cardiac cycles along a specific plane [13, 14]. The use of M-mode in conjunction with ECG allows LV measurements to be made more reliably (see **Figure 1** for examples). The LV end-diastolic internal diameter (LVIDd) should be measured during the onset of the QRS complex and near the end of the T wave for LV internal dimension during systole (LVIDs) [15]. Comparisons are ideally made against breed-specific LV measurements, but where this is not possible values should be compared to breeds of a similar size and weight [9].

Fractional shortening is a major indicator of systolic function, where values less than 20–25% suggest impaired contractility. This is calculated as follows:

$$FS = \left[(LVIDd - LVIDs) / LVIDd\right] X\,100 \tag{1}$$

In some cases, fractional shortening may be misleading, for example in athletic breeds values may appear to be lower, and in dogs with severe mitral regurgitation they can be higher [2, 15, 16], therefore caution is recommended when making a diagnosis. End point to septal separation (EPSS) is the minimum distance between the anterior mitral leaflet (E point) to the LV septal wall, during the rapid filling phase of diastole [17, 18]. An EPSS measurement of >10 mm in any breed is considered abnormal. Increased values can occur in volume overload or reduced fractional shortening resulting in LV dilation [11, 13].

A characteristic feature of DCM is the ventricular chamber becoming more spherical as the ventricle dilates [19]. The index of sphericity (SI) is calculated during diastole by dividing the LV length with the LV width, where a value <1.65 is indicative of an abnormally rounder chamber [10].

Although M-mode echocardiography is commonly used (**Figure 1**), its one-dimensional nature restricts the spatial information provided and the technique is reliant on geometrical assumptions that may be altered during pathological states [8, 14, 20, 21]. The American Society of Echocardiography recommends that linear measurements should not be used to calculate LV volume, and suggest that the biplanar Simpson's methods of discs (SMOD) is more suitable [22]. A study in Dobermans concluded that SMOD is more sensitive than M-mode in detecting early echocardiographic changes observed in DCM. Thus, it is recommended that where possible, SMOD reference values should be used [8, 10].

The SMOD volume formula is based on tracing the endocardial border across the mitral annulus and measuring the long axis of the left ventricle. The LV cavity is divided into 20 discs of equal height, where the cross-sectional area of each disc is based on the diameters obtained from two orthogonal LV views. End-diastolic (EDV) and end-systolic (ESV) volumes are calculated from the summation of the stacked discs, often utilising the software available on many ultrasound machines. These volumes can be normalised to body surface area to give volume indices (EDV-I and ESV-I). The current recommendations are that an ESV-I > 80 ml/m^2 is a strong indication of systolic dysfunction [2, 8, 22]. Ejection fraction (EF) is calculated in a similar way to fractional shortening, but volume measurements are used as follows:

$$EF = (EDV - ESV) / EDV \tag{2}$$

Thus, EF takes into account both radial and longitudinal cardiac changes, where dogs with EF <40% are reflective of reduced inotropy [2, 22].

ECG is used for the detection of arrhythmias but, as arrhythmias are often intermittent, they may be missed by an in-house 5-minute ECG commonly used in first opinion practice. 24-hour ambulatory ECG (Holter) monitoring provides a more representative assessment, where the patient is not restricted to a clinical setting, yet continuous monitoring is enabled [11, 23]. Holter monitoring for 24 hours is often used to identify individuals that have arrhythmias, with ventricular arrhythmias being common in animals [88]. Owners additionally record activities such as sleeping or running to enable changes in heart rate or rhythm to be correlated with patient activity [24]. Most dogs with DCM show evidence of arrhythmias and in certain breeds these may precede echocardiographic abnormalities [25, 26].

Atrial fibrillation (AF) is a common supraventricular tachyarrhythmia in large and giant breeds [3, 27]. Despite the presence of AF in a large percentage of dogs with DCM, the mechanistic and clinical relationship between DCM and AF has not been clarified [16–19]. **Figure 1A** shows an example of a six lead ECG showing fast atrial fibrillation in a dog with DCM. One canine study showed that 80.5% of individuals with DCM also had a diagnosis of AF [20], with another study showing occurrence in 87.6% of the patients [19]. Most individuals were diagnosed with AF at the same time as DCM or in the 2 years prior to the diagnosis of DCM, which indicates that AF may be a precursor to a clinical diagnosis of DCM [20]. Therefore, individuals diagnosed with AF should be carefully monitored and regularly presented for heart testing to check for DCM. There is the potential to improve the survival of individuals diagnosed with AF by treating them with drugs such as pimobendan prior to the development of DCM or heart failure [21].

Ventricular premature complexes (VPC) appear to be more common in the Doberman and Boxer breeds than other breeds. In Dobermans, >300 VPCs/24 hours, or two successive Holter recordings within one year showing between 50 and 300 VPCs/24 hours is considered diagnostic for preclinical DCM, even if echocardiographic findings appear normal [10]. Thus, Holter monitoring is useful in identifying Dobermans that are destined to develop DCM. Similar diagnostic reference ranges are lacking in the Boxer but >50 VPCs/24 hours would be considered to be abnormal [2].

The European Society of Veterinary Cardiology (ESVC) has proposed a scoring system to aid in the diagnosis of DCM, especially for dogs that present with equivocal findings. The following cardiac changes fall into the major criteria and are allocated 3 points each: (i) LV enlargement, (ii) reduced systolic function, and (iii) increased LV sphericity. The remaining findings are considered minor (1 point each): (a) arrythmias in specific breeds; (b) AF; (c) increased EPSS; (d) increased pre-ejection period: ejection time ratio; (e) LV fractional shortening in equivocal range; (f) left or biatrial enlargement. A total of ≥6 points is indicative of DCM and a score of 1–5 should encourage repeated examination for evidence of disease progression [2].

Annual screening using echocardiography and Holter monitoring, has been recommended for breeds genetically predisposed to DCM, including Dobermans, Boxers, Newfoundlands, Great Danes and Irish Wolfhounds (IWH). Detection of the pre-clinical phase allows earlier therapeutic intervention, can improve prognosis, and enables the removal of affected dogs from breeding programmes if appropriate [10, 28, 29]. This is particularly important in Dobermans as 30% die suddenly prior to the onset of congestive heart failure [30, 31]. However, yearly testing can be expensive as it often requires referral to specialists thus restricting accessibility. In some countries, breed groups/welfare groups have set up testing programmes, and some even support these financially or fundraise, due to the concern about the numbers of animals developing cardiovascular problems.

Given the complex nature of diagnosis, that gold standard tests may not always be available for every client, and the financial restraints faced by some owners, development of further and/or potentially cost effective diagnosis tools are always needed [5, 7]. Biomarkers such as N-Terminal pro B-type natriuretic peptide (NT-proBNP) have been used in humans to identify patients with occult LV dysfunction. The NT-proBNP assay is useful to differentiate between cardiac and non-cardiac causes of respiratory distress, where conventional testing alone could lead to ambiguous results [28, 32]. In a study of 328 Dobermans, those that had plasma concentrations of NT-proBNP >400 pmol/L, in the absence of renal dysfunction, were more likely to have echocardiographic abnormalities. However, the results cannot be considered diagnostic as NT-proBNP concentrations overlapped in groups of dogs with and without preclinical DCM [7, 33]. The use of in-house 5-minute ECG has reasonable specificity and in Dobermans the detection of 1 VPC strongly suggests that >100 VPC would be recorded via Holter. However, due to poor sensitivity, absence of VPC should not rule out the possibility of DCM [6, 23]. The emphasis is that the results from these tests should not be used to establish a diagnosis, but rather to identify dogs that would benefit further from more costly diagnostic tests [34].

Two histopathological variations of canine DCM have been identified: “attenuated wavy fibre type” and “fatty infiltration type” [22] indicating that differing types of canine DCM exist. The fatty infiltration type has only been reported in Doberman Pinschers, Estrela mountain dogs, Great Danes, and Boxers [22–25]; whereas the wavy fibre type can occur in all breeds [22, 23]. As the wavy fibre type is found across breeds, and in many individuals, it could be the tissue’s response to the other processes of DCM. In general, atrophy, or attenuation, of muscle fibres is often a result of processes that prevent normal contractile ability: contractile ability is consistently compromised in DCM [26]. The clinical relevance and prevalence of these two histopathological variants remain to be established, and as post mortem tissue is required presently for phenotypic analysis this may not be useful in a clinical setting but could provide valuable research insights into the disorder [27].

The long-term prognosis of canine DCM can be highly variable, with well managed dogs maintaining a good quality of life for many years and others dying within weeks of diagnosis despite careful clinical management [19, 28, 29]. There are some breed specific prognosis trends such as Doberman Pinschers which generally have a poor prognosis. Their mean time to death (from diagnosis) is in the range of 7.4 to 9.7 weeks [29, 30], which is low in comparison to other breeds reported to be about four times that at 34 weeks [29]. Doberman Pinschers also had the lowest upper quartile range for survival time in a study, but analysis carried out by the same research showed that Great Danes also suffer from a poor prognosis, with the lowest median survival time of any breed [17, 28].

Age of onset can also affect prognosis and may be a useful indicator that differing types of canine DCM exist. Portuguese water dogs have a specific juvenile form of DCM, where age of onset is measured in weeks from birth [31, 32], while in most other cases age of onset is measured in years [17]. Great Danes have a mean age of onset of 4.8 (SD ± 2.3) years [33], comparable to Irish Wolfhound mean age of onset of 4.40 (SD ± 2.03) years [34]; but lower than Doberman Pinscher’s at 7.3 years in males and 8.6 years in females [30]. Once identified, knowledge about the differing canine DCM types could benefit current and future potential treatments in addition to elucidating other clinically important factors in canine DCM, such as longevity and prognosis.

3. Treatment and management options

The ultimate aim of treatment is to cure disease, but this is currently not always possible. When a cure is not available, treatment of disease is aimed at reducing the impact of the disease, extending lifespan, and maintaining quality of life. Treatment of DCM in people is focused primarily on managing symptoms if at the overt stage of disease or, if presented with a preclinical case, prolonging the time between diagnosis and congestive heart failure [35]. Due to the predisposition of certain canine breeds, preclinical cardiac screening can help diagnose early abnormal findings, leading to a more successful diagnosis and potentially a management and treatment regime [36], in addition to possibly altering breeding programmes and preferences for some owners/breeders.

As with human DCM, in veterinary cases the ultimate aim is to minimise the effect of heart failure with attempts to delay disease progression depending on the stage of diagnosis [37]. Strategic treatments presently include vasodilators, angiotensin-converting enzyme (ACE) inhibitors, diuretics and positive inotropes. With atypical breeds or American cocker spaniels and Golden retrievers, dietary supplementation of taurine and L-carnitine is usually recommended if the suspected aetiology of DCM is diet-related [38–40].

In regards to treating earlier stages of DCM, there is often a focus on the prevention of further myocardial dysfunction by using the cardioprotective effect of ACE inhibitors [41]. These interventions focus on the vasodilation of blood vessels by reducing angiotensin II effects within the renin-angiotensin-aldosterone system (RAAS), a system heavily associated with severe heart disease through aldosterone release [42]. ACE inhibitors currently approved for veterinary purposes include benazepril, enalapril, imidapril and Ramipril, with the former being used to treat initial stages of heart failure and the latter three for more progressive stages [43].

Furosemide and spironolactone are both recommended diuretics, with the former inhibiting the re-absorption of sodium and chloride in the thick ascending loop of Henle to promote natriuresis. The latter is a weak potassium-sparing diuretic which works as an aldosterone antagonist. Monotherapy of these diuretics are not recommended [44] as they reduce plasma volume, further stimulating RAAS activity [42]. Therefore, diuretics are often used in combination with ACE inhibitors.

In Ref. to the undetermined link between canine DCM and atrial fibrillation (AF), for these patients, positive inotropes such as digoxin are particularly effective by increasing vagal tone, hence, decreasing heart rate [45]. Another drug often used is diltiazem, a calcium channel blocker also slowing conduction through the atrioventricular node. Although it is in fact a negative inotrope, it is particularly effective in dogs with AF by increasing diastolic filling time and therefore improving cardiac output [46]. Even though both of these drugs can be used individually, research indicates that using these drugs in combination can prove to be more effective [47]. With digoxin having a narrow therapeutic index, digoxin toxicity can occur and gradual withdrawal should take place if clinical signs such as depression, anorexia and vomiting are seen [45]. In addition, pimobendan has been investigated in relation to prolonging the onset of coronary heart failure due to DCM by having similar effects to digoxin whilst advantageously keeping myocardial oxygen demand to a minimum [42]. Pimobendan has also been shown to increase the survival rate in Dobermans [36] and we can assume it offers similar benefits to other breeds.

Although the majority of canine DCM treatments focus on managing the condition whilst optimising quality of life, diet-associated DCM can be potentially reversible with certain supplements such as taurine and L-carnitine [40]. Taurine deficiency is a prime cause of DCM in cats, evidence suggests micronutrient shortages (e.g. selenium, iron) are linked to human DCM, and there are some trials suggesting there may be a link in certain canine breeds, including American cocker spaniels, Golden retrievers [38, 40]. Accumulating research indicates that nutrient imbalances may cause or exacerbate DCM due to reduced myocardial expenditure [48]. There are minimal to no side-effects of taurine supplementation in dogs [49], therefore treating American cocker spaniels, Golden retrievers and atypical breeds presenting with DCM with taurine pending the results of blood testing may be prudent. Furthermore, alongside nutraceuticals such as taurine and L-carnitine, omega-3 fatty acids have been shown to significantly reduce muscle loss and prostaglandin E2 production, and these results also correlate with increased survival rate [50]. However, these supplementary products are very expensive, especially when prescribed for larger breed dogs, so finding the possible aetiology for the disease is vital to determine the efficacy of these products [49].

Beta-blockers such as carvedilol have shown to have positive impacts for people with DCM [51–53]. Although carvedilol is a popular choice in humans with DCM and congestive heart failure [54, 55], there is limited evidence to date that these are beneficial in canine breeds with DCM [56]. Where used, careful monitoring should be ensured and treatment is best tolerated in dogs during the early stages of DCM [42].

In latter stages of DCM, the last resort for treating human heart disease includes transplants or inserting cardiac assist devices such as pacemakers [37], but these are far less realistic for canine patients due to cost, disease prognosis and surgical risk for the animal. Therefore, medical management based on symptomatic relief is the only current option for these patients [37].

Given that identifying, preventing, managing, and treating canine DCM is still a clinical challenge, new approaches are still needed. One new approach focusses on therapeutic gene transfer to target underlying molecular defects of ventricular dysfunction. Gene therapy could replace a defective gene, or part of a gene, with a functional one, resulting in effective amounts of a certain protein being produced that was once deficient [57, 58]. Evidence suggests Ca^{2+} handling is damaged in a heart with cardiac failure, and experiments involving altering calcium-handling protein levels in rodents through cardiac gene therapy have been successful [59]. Although achieving myocardial transduction in larger animal models presents many difficulties, therapeutic gene transfer may be a viable option in the future to treat canine DCM regardless of underlying cause. Another example of gene therapy is the use of vascular endothelial growth factor-B167 (VEGF-B167), which was administered via an intracoronary route in a study of 10 dogs with DCM [60]. The method was well tolerated by the canine patients. VEGF-B has anti-apoptotic and cardioprotective effects, so could be used to mitigate progression of DCM. Naturally many genes are associated with cardiomyopathy or in the drug and gene pathways [27, 61, 62], and additionally multiple genetic associations have been associated with the disorder [20] which potentially increases the complexity of gene therapy interventions.

Another of the requirements of gene therapy is understanding which mutations cause DCM in which species/breeds, therefore large trials are necessary in order to understand the basics, in addition to understanding the best methods of introducing gene therapy itself. For example gene therapy trials into dystrophin delivery via viral vectors has been promising but cardiovascular tissue is more complex than skeletal muscle, even when delivered via intravenous injection in dogs [63, 64]. Therefore gene therapy may be a very promising avenue of research, but the impact of gene associations and delivery methods may need to be taken into account for differing individuals and/or breeds.

Other new approaches being researched currently include myocardial regeneration therapies such as the use of resident cardiac stem cells, bone marrow stem cells, and skeletal myoblasts via transplantation therapies [65]. Cellular cardiomyoplasty has been explored via transplants of skeletal muscle cells into myocardium of canine DCM hearts. In early trials three of five dogs died early of tachyarrhythmias and pulmonary embolism, however two dogs survived and showed improved function of the heart [66]. Further investigation into this method is needed to understand the mechanisms and to prove if it can be successfully used in the clinic. Previous research on hamsters using smooth muscle cell transplants also showed clinical improvement in heart function [67], therefore this method could be developed for dogs and other species.

The use of cardiosphere-derived stem cells (CDCs) from adult dog hearts has also been trialled [68]. These cells can differentiate into cardiomyocytes *in vitro* and a study in Doberman pinschers diagnosed with DCM detailed patients given an infusion of canine CDCs via the coronary vessels [68, 69]. There was no rejection of the cells, and no adverse reactions were reported [69]. It has been suggested that therapy using CDCs can slow down ventricular enlargement and the effects of DCM such as systolic dysfunction. Studies on larger sample sizes need to be undertaken.

2-deoxyadenosine triphosphate (dATP) is an energy substrate that can be used by myosin instead of ATP during formation of cross-bridges in myocardial contraction. A study discovered that substituting ATP with dATP in both normal and DCM hearts leads to increased activity of myofilaments and increased systolic function [70]. Another potential pharmaceutical method is active myosin using drugs such as danicamtiv, which has undergone in vitro and in vivo trials in dogs, people and other mammals [71].

Transvenous electrical cardioversion of AF involves the placement of electrode coils in the vasculature of the heart, with the right atrium acting as the anode, and the left pulmonary artery acting as the cathode. A connection with a defibrillator is made and electrical shocks are given in time with the R wave. In a study of two canine cases affected by DCM, treatment resulted in the heartbeat returned to a normal sinus rhythm [72]. There is therefore potential for this therapy to be used to reduce the risk of AF leading to DCM development.

4. Genetics of cardiomyopathy and association with other conditions

Despite known pre-dispositions to diseases in the majority of UK kennel club registered breeds, there are currently only 93 disease associated variants identified and only 61 genetic tests commercially available across all breeds [73–75]. In human genetic testing there are tests available for 10485 conditions [76, 77]. This is a large difference despite close similarity between many diseases affecting dogs and humans [78–80]. Many tools have been utilised when identifying genetic variants associated with human diseases, these include genome wide association studies, candidate gene studies, whole exome sequencing, and whole genome sequencing [81–84]. These tools have also been utilised to some extent in some canine studies, but these approaches are often restrictively expensive for companion animal studies with limited funding available [85, 86]. The cost of these methods is delaying their clinical and research application in evaluating the genetic basis of canine diseases. Identification of susceptibility loci for these diseases has the potential to bring improvements in diagnosis and may lead to improved treatments in line with diseases which already have genetic loci associated with them [87–92].

The development and progression of common non-communicable diseases such as heart disease are influenced by a combination of risk factors. It has been shown in several diseases that interactions between environmental and genetic risk factors are important in the development and progression of disease [13, 15]. Environmental risk factors for people with DCM are often modifiable as individuals can make informed choices with regards to lifestyle changes to reduce their risk of developing disease [7, 10, 93], likewise dog owners may be able to modify environmental risk factors for their animals. Currently genetic risk factors are not modifiable, although genome editing technologies may allow this in the future [94]. Despite this, individuals identified as having a high genetic risk of developing a disease are currently able to reduce their risk of developing disease. Options available include making lifestyle changes to reduce other risk factors, increased disease surveillance, and prophylactic medicine and surgery [4, 7, 10, 92, 93, 95, 96]. Increased disease surveillance can allow for early disease detection and therefore early treatment, which is associated with improved prognosis [87, 92]. An additional benefit of knowledge of the underlying genetic cause of disease is that it could lead to targeted treatments such as rectifying the defective gene or drugs targeting affected pathways [88, 89].

There are over 50 genes associated with DCM in people, some with multiple mutations, whereas there have only been 10 loci associated with canine DCM [85, 86, 97–100]. We have recently reviewed the genetics of canine and human DCM [27]. There has also been a RNAseq study examining the difference in expression of genes between canine DCM hearts and non-DCM hearts [101]. In the RNAseq study, genes involved in cellular energy metabolism were expressed less in the DCM hearts than the non-DCM hearts [101]. In several breeds canine DCM has been shown to be heritable [29, 31, 34]. To date mutations in only two genes (*PDK4* and *STRN*) and a single nucleotide polymorphism (SNP) on chromosome 5 have been associated with canine DCM [85, 97, 98, 102], and these are limited to a few breeds, suggesting additional genetic causes remain unknown. Of note are also cardiac troponin T and dystrophin which have both been highlighted in people, dogs and other species in relation to dilated cardiomyopathy [103–105], as mentioned in relation to gene therapy trials.

Genetic models and studies have also shown sex-linked genetic influences in relation to pathogenesis and a multigenic contribution to canine DCM [102]. The work showed that by combining three factors (PDK4, Chr5 TIGRP2P73097 SNP and an X-linked locus) DCM incidence could be more accurately predicted in a canine population. Overall this data showed that models incorporating multiple factors were more effective than those incorporating a single factor [102]. This has implications for future studies of the genetics and management of DCM, including monitoring which could enable earlier clinical intervention of individuals who are high risk.

In addition to commonly being diagnosed with DCM, IWHs are frequently diagnosed with atrial fibrillation (AF) [16, 19]. Despite the presence of AF in a large percentage of dogs with DCM, the mechanistic and clinical relationship between DCM and AF has not been clarified [16–19]. IWHs can develop DCM without AF, though it seems that <2% of IWHs with AF do not go on to develop DCM [16, 19]. If AF is a potential precursor to DCM, the time from diagnosis of AF to DCM is important, as if it is several years than the presence of AF could be less of a concern than if it is merely a few months. Also, if AF is a precursor to DCM, there is the potential to give individuals diagnosed with AF drugs such as the phosphodiesterase III/calcium sensitizing drug, pimobendan, to improve survival [21]. In addition to the potential clinical implications of AF diagnosis, if AF can be shown to be related to DCM, both diagnoses can be used for genetic association testing. This also has implications for individuals included in the unaffected group for genetic association testing, as individuals included in the unaffected group must be free of both DCM and AF. There is some evidence that males are affected by DCM more often, or earlier in life than females [16, 18, 20, 30, 106].

Hypothyroidism has also been linked to increased DCM rates in some studies. For example Doberman Pinchsers with DCM were 2.26 times more likely to have or develop hypothyroidism [107], and suggestions to links between the two conditions was also proposed in two Great Dane individuals [108]. Some research has not necessarily shown a link between the two disorders [109–111], therefore there are still questions around links, especially causal ones. Although exact mechanisms, or indeed associations have yet to be determined, evidence does support that the thyroid hormones can have positive inotropic and chronotropic effects, and that under both experimental and in patients these may have important influences over cardiovascular health and functions [112].

5. Canines as a model for human cardiomyopathy

Current understanding of disease processes and treatments is based on studying affected individuals compared to unaffected individuals, along with the use of animal models of disease, cell lines, and computer simulations [113–119]. Natural models of disease allow researchers access to additional cases of disease without inducing disease and causing additional suffering, because the animals involved develop disease irrespective of involvement in a study. Therefore, a relevant resource for investigating health and disease is the companion animal population, within which dogs in particular are useful as natural models of the equivalent human disease [1, 120–123].

The canine population overall is genetically heterogeneous, yet breeds are comparatively homogeneous which enhances their value as genetic models of disease [124]. Each breed of dog is a closed population and ancestry can typically be traced for many generations, often to the founding members of the breed [124–126]. This facilitates understanding the mode of inheritance of traits and diseases, and also restricts the amount of genetic diversity within a breed [34, 121, 125, 127]. Founder effects and subsequent inbreeding within pedigree dog breeds have led to differing allele frequencies between breeds, and some breeds are more prone to developing particular conditions than others [124, 128, 129]. This makes breeds with homologs of human conditions ideal for identifying potential genetic loci associated with disease for both canine and human benefit [123, 130].

Many canine disease phenotypes can be closely matched to human disease phenotypes with similar disease progression, pathology, treatment options, and prognosis [78–80]. Indeed there are currently 383 potential canine models for human disease listed in OMIA (Online Mendelian Inheritance in Animals), greater than any other species [78, 79]. Dogs are typically treated as family members and so inhabit the same environment as their owners with the associated exposure to the same potential environmental toxins, including, for example, air pollution [131]. Pet dogs also frequently benefit from high quality medical care, such that illnesses are detected and treated promptly, similar to the human population [132]. These characteristics of the canine population make it a valuable resource as a model of human disease. Examples of diseases with homologies in humans and dogs include diabetes, cardiomyopathies, cancers, and eye diseases [79, 120, 123, 133, 134].

6. Conclusions

Diagnosis of disease is informed by patient symptoms, family history, medical testing, and in some cases genetic testing [135–137]. With greater understanding of disease progression, these aspects can be more accurately assessed and give earlier accurate diagnoses [87]. Early diagnosis can enable early treatment, which can result in improved outcomes compared to patients diagnosed later in the disease course [87, 92, 138]. Screening for disease prior to the onset of symptoms can catch diseases at an early stage, but may not be useful or affordable for an entire population. Screening asymptomatic individuals can be recommended when there is an additional reason to suspect that an individual may develop disease, such as family/breed history of the disease or a genetic test result indicating susceptibility to disease [95, 96]. A positive genetic test result often does not fully predict disease development, but merely indicates that the individual has a genetic pre-disposition to developing the disease [139]. Thus, a positive genetic test result does not always result in direct medical intervention, yet it can lead to increased awareness of the disease and enrolment of the individual on a health monitoring programme [92, 96]. Not only can the dog be an excellent model, but lessons can also be learned from other species with DCM [61].

Acknowledgements

The authors gratefully acknowledge generous funding from the BBSRC University of Nottingham Doctoral Training Programme BB/J014508/1 and the School of Veterinary Medicine and Science. We would also like to thank The Wellcome Trust, The Academy of Medical Sciences for INSPIRE research funding.
Catrin Rutland https://orcid.org/0000-0002-2009-4898.

Conflict of interest

The authors declare no conflict of interest.

Abbreviations

ACE	Angiotensin-converting enzyme
AF	Atrial fibrillation
CDCs	Cardiosphere-derived stem cells
dATP	Deoxyadenosine triphosphate
DCM	Dilated cardiomyopathy
EF	Ejection fraction
ECG	Electrocardiograph
EDV	End-diastolic volume
EDV-I	End-diastolic volume indices
ESV	End-systolic volume
ESV-I)	End-systolic volumes indices
EPSS	End point to septal separation
ESVC	European Society of Veterinary Cardiology
SI	Index of sphericity
IWH	Irish Wolfhounds
LV	Left ventricular
LVIDd	Left ventricular end-diastolic internal diameter
LVIDs	Left ventricular internal dimension during systole
NT-proBNP	N-Terminal pro B-type natriuretic peptide
OMIA	Online Mendelian Inheritance in Animals

RAAS	Renin-angiotensin-aldosterone system
SMOD	Simpson's methods of discs
VEGF-B167	Vascular endothelial growth factor-B167
VPC	Ventricular premature complexes

Author details

Siobhan Simpson[1], Kara-Zoë Kordtomeikel[1], Serena Wong[1], Samantha Bennison[1], Samir A.A. El-Gendy[2], Malcolm Cobb[1] and Catrin Sian Rutland[1*]

1 School of Veterinary Medicine and Science, University of Nottingham, Nottingham, UK

2 Department of Anatomy and Embryology, Faculty of Veterinary Medicine, Alexandria University, Egypt

*Address all correspondence to: catrin.rutland@nottingham.ac.uk

References

[1] Egenvall A, Bonnett BN, Häggström J. Heart Disease as a Cause of Death in Insured Swedish Dogs Younger Than 10 Years of Age. Journal of Veterinary Internal Medicine. 2006;20(4):894-903.

[2] McNally EM, Golbus JR, Puckelwartz MJ. Genetic mutations and mechanisms in dilated cardiomyopathy. The Journal of Clinical Investigation. 2013;123(1):19-26.

[3] Banks E, Joshy G, Weber M, Liu B, Grenfell R, Egger S, et al. Tobacco smoking and all-cause mortality in a large Australian cohort study: findings from a mature epidemic with current low smoking prevalence. BMC Medicine. 2015;13(1):38.

[4] Domchek SM, Friebel TM, Singer CF, et al. Association of risk-reducing surgery in brca1 or brca2 mutation carriers with cancer risk and mortality. The Journal of the American Medical Association. 2010;304(9):967-75.

[5] McFadden E, Stevens R, Glasziou P, Perera R. Implications of lower risk thresholds for statin treatment in primary prevention: Analysis of CPRD and simulation modelling of annual cholesterol monitoring. Preventive Medicine. 2015;70:14-6.

[6] Hall JM, Lee MK, Newman B, Morrow JE, Anderson LA, Huey B, et al. Linkage of early-onset familial breast cancer to chromosome 17q21. Science. 1990;250(4988):1684-9.

[7] Parkin DM. 3. Cancers attributable to consumption of alcohol in the UK in 2010. British Journal of Cancer. 2011;105(S2):S14-S8.

[8] Parkin DM. 2. Tobacco-attributable cancer burden in the UK in 2010. British Journal of Cancer. 2011;105(S2):S6-S13.

[9] Parkin DM. 5. Cancers attributable to dietary factors in the UK in 2010. British Journal of Cancer. 2011;105(s2):s24-S6.

[10] Parkin DM. 9. Cancers attributable to inadequate physical exercise in the UK in 2010. British Journal of Cancer. 2011;105(S2):S38-S41.

[11] Wooster R, Bignell G, Lancaster J, Swift S, Seal S, Mangion J, et al. Identification of the breast cancer susceptibility gene BRCA2. Nature. 1995;378(6559):789-92.

[12] Chu Y-H, Tzeng S-L, Lin C-W, Chien M-H, Chen M-K, Yang S-F. Impacts of MicroRNA Gene Polymorphisms on the Susceptibility of Environmental Factors Leading to Carcinogenesis in Oral Cancer. PLoS ONE. 2012;7(6):e39777.

[13] Nickels S, Truong T, Hein R, Stevens K, Buck K, Behrens S, et al. Evidence of Gene–Environment Interactions between Common Breast Cancer Susceptibility Loci and Established Environmental Risk Factors. PLoS Genetics. 2013;9(3): e1003284.

[14] Sharafeldin N, Slattery ML, Liu Q, Franco-Villalobos C, Caan BJ, Potter JD, et al. A Candidate-Pathway Approach to Identify Gene-Environment Interactions: Analyses of Colon Cancer Risk and Survival. Journal of the National Cancer Institute. 2015;107(9):djv160.

[15] Song C, Chang Z, Magnusson PKE, Ingelsson E, Pedersen NL. Genetic factors may play a prominent role in the development of coronary heart disease dependent on important environmental factors. Journal of Internal Medicine. 2014;275(6):631-9.

[16] Brownlie SE, Cobb MA. Observations on the development of congestive heart failure in Irish

wolfhounds with dilated cardiomyopathy. Journal of Small Animal Practice. 1999;40(8):371-7.

[17] Martin MW, Stafford Johnson MJ, Celona B. Canine dilated cardiomyo pathy: a retrospective study of signalment, presentation and clinical findings in 369 cases. Journal of Small Animal Practice. 2009;50(1):23-9.

[18] Tidholm A, Jonsson L. A retrospective study of canine dilated cardiomyopathy (189 cases). Journal of American Animal Hospital Association. 1997;33(6):544-50.

[19] Vollmar AC. The prevalence of cardiomyopathy in the Irish wolfhound: a clinical study of 500 dogs. Journal of the American Animal Hospital Association. 2000;36(2):125-32.

[20] Simpson S, Dunning MD, Brownlie S, Patel J, Godden M, Cobb M, et al. Multiple Genetic Associations with Irish Wolfhound Dilated Cardiomyopathy. Biomed Res Int. 2016;2016:6374082.

[21] Vollmar AC, Fox PR. Long-term Outcome of Irish Wolfhound Dogs with Preclinical Cardiomyopathy, Atrial Fibrillation, or Both Treated with Pimobendan, Benazepril Hydrochloride, or Methyldigoxin Monotherapy. Journal of Veterinary Internal Medicine. 2016;30(2):553-9.

[22] Tidholm A, Jonsson L. Histologic characterization of canine dilated cardiomyopathy. Veterinary Pathology. 2005;42(1):1-8.

[23] Aupperle H, Marz I, Baldauf K, Roggon N, Kresken JG. Pathology of DCM in Great Danes. Journal of Veterinary Internal Medicine. 2014;28(2):730-.

[24] Lobo L, Carvalheira J, Canada N, Bussadori C, Gomes JL, Faustino AM. Histologic characterization of dilated cardiomyopathy in Estrela mountain dogs. Veterinary Pathology. 2010;47(4):637-42.

[25] Stephenson HM, Fonfara S, López-Alvarez J, Cripps P, Dukes-McEwan J. Screening for dilated cardiomyopathy in Great Danes in the United Kingdom. Journal of veterinary internal medicine / American College of Veterinary Internal Medicine. 2012;26(5):1140-7.

[26] Tidholm A, Haggstrom J, Borgarelli M, Tarducci A. Canine idiopathic dilated cardiomyopathy. Part I: Aetiology, clinical characteristics, epidemiology and pathology. The Veterinary Journal. 2001;162(2):92-107.

[27] Simpson S, Edwards J, Ferguson-Mignan TFN, Cobb M, Mongan NP, Rutland CS. Genetics of Human and Canine Dilated Cardiomyopathy. International Journal of Genomics. 2015;2015:13.

[28] Martin MWS, Stafford Johnson MJ, Strehlau G, King JN. Canine dilated cardiomyopathy: a retrospective study of prognostic findings in 367 clinical cases. Journal of Small Animal Practice. 2010;51(8):428-36.

[29] Petric AD, Stabej P, Zemva A. Dilated cardiomyopathy in Doberman Pinschers: Survival, Causes of Death and a Pedigree Review in a Related Line. Journal of Veterinary Cardiology. 2002;4(1):17-24.

[30] Calvert CA, Pickus CW, Jacobs GJ, Brown J. Signalment, Survival, and Prognostic Factors in Doberman Pinschers With End-Stage Cardiomyopathy. Journal of Veterinary Internal Medicine. 1997;11(6):323-6.

[31] Dambach DM, Lannon A, Sleeper MM, Buchanan J. Familial dilated cardiomyopathy of young Portuguese water dogs. Journal of veterinary internal medicine / American

College of Veterinary Internal Medicine. 1999;13(1):65-71.

[32] Werner P, Raducha MG, Prociuk U, Sleeper MM, Van Winkle TJ, Henthorn PS. A novel locus for dilated cardiomyopathy maps to canine chromosome 8. Genomics. 2008;91(6):517-21.

[33] Meurs KM, Miller MW, Wright NA. Clinical features of dilated cardiomyopathy in Great Danes and results of a pedigree analysis: 17 cases (1990-2000). Journal of the American Veterianry Medical Association. 2001;218(5):729-32.

[34] Distl O, Vollmar AC, Broschk C, Hamann H, Fox PR. Complex segregation analysis of dilated cardiomyopathy (DCM) in Irish wolfhounds. Heredity. 2007;99(4):460-5.

[35] Guttmann OP, Mohiddin SA, Elliott PM. Almanac 2014: cardiomyopathies. Heart. 2014;100(10):756-64.

[36] Summerfield NJ, Boswood A, O'Grady MR, Gordon SG, Dukes-McEwan J, Oyama MA, et al. Efficacy of pimobendan in the prevention of congestive heart failure or sudden death in Doberman Pinschers with preclinical dilated cardiomyopathy (the PROTECT Study). J Vet Intern Med. 2012;26(6):1337-49.

[37] Sleeper MM, Bish LT, Sweeney HL. Status of therapeutic gene transfer to treat canine dilated cardiomyopathy in dogs. Vet Clin North Am Small Anim Pract. 2010;40(4):717-24.

[38] Kaplan JL, Stern JA, Fascetti AJ, Larsen JA, Skolnik H, Peddle GD, et al. Correction: Taurine deficiency and dilated cardiomyopathy in golden retrievers fed commercial diets. PLoS One. 2018;13(12):e0210233.

[39] Kramer GA, Kittleson MD, Fox PR, Lewis J, Pion PD. Plasma taurine concentrations in normal dogs and in dogs with heart disease. J Vet Intern Med. 1995;9(4):253-8.

[40] Kittleson MD, Keene B, Pion PD, Loyer CG. Results of the multicenter spaniel trial (MUST): taurine- and carnitine-responsive dilated cardiomyopathy in American cocker spaniels with decreased plasma taurine concentration. J Vet Intern Med. 1997;11(4):204-11.

[41] O'Grady MR, O'Sullivan ML, Minors SL, Horne R. Efficacy of benazepril hydrochloride to delay the progression of occult dilated cardiomyopathy in Doberman Pinschers. J Vet Intern Med. 2009;23(5):977-83.

[42] Oyama MA. Cardiac Drugs for Treatment of Canine Heart Failure. NAVC Clinician's Brief. 2009;7:56-9.

[43] Lefebvre HP, Brown SA, Chetboul V, King JN, Pouchelon JL, Toutain PL. Angiotensin-converting enzyme inhibitors in veterinary medicine. Curr Pharm Des. 2007;13(13):1347-61.

[44] O'Grady MR, O'Sullivan ML. Dilated cardiomyopathy: an update. Vet Clin North Am Small Anim Pract. 2004;34(5):1187-207.

[45] Merrett D. Digoxin therapy. Aust Vet J. 2000;78(9):612-5.

[46] Stephenson H. Dilated cardiomyopathy therapy. Vet Times. 2011.

[47] Gelzer AR, Kraus MS, Rishniw M, Moise NS, Pariaut R, Jesty SA, et al. Combination therapy with digoxin and diltiazem controls ventricular rate in chronic atrial fibrillation in dogs better than digoxin or diltiazem monotherapy: a randomized crossover study in 18

dogs. J Vet Intern Med. 2009;23(3): 499-508.

[48] Adin D, DeFrancesco TC, Keene B, Tou S, Meurs K, Atkins C, et al. Echocardiographic phenotype of canine dilated cardiomyopathy differs based on diet type. J Vet Cardiol. 2019;21:1-9.

[49] Sanderson SL. Taurine and carnitine in canine cardiomyopathy. Vet Clin North Am Small Anim Pract. 2006;36(6):1325-43, vii-viii.

[50] Freeman LM, Rush JE, Kehayias JJ, Ross JN, Jr., Meydani SN, Brown DJ, et al. Nutritional alterations and the effect of fish oil supplementation in dogs with heart failure. J Vet Intern Med. 1998;12(6):440-8.

[51] Packer M, Bristow MR, Cohn JN, Colucci WS, Fowler MB, Gilbert EM, et al. The effect of carvedilol on morbidity and mortality in patients with chronic heart failure. U.S. Carvedilol Heart Failure Study Group. N Engl J Med. 1996;334(21):1349-55.

[52] Packer M, Coats AJ, Fowler MB, Katus HA, Krum H, Mohacsi P, et al. Effect of carvedilol on survival in severe chronic heart failure. N Engl J Med. 2001;344(22):1651-8.

[53] Poole-Wilson PA, Swedberg K, Cleland JG, Di Lenarda A, Hanrath P, Komajda M, et al. Comparison of carvedilol and metoprolol on clinical outcomes in patients with chronic heart failure in the Carvedilol Or Metoprolol European Trial (COMET): randomised controlled trial. Lancet. 2003;362(9377):7-13.

[54] Kveiborg B, Major-Petersen A, Christiansen B, Torp-Pedersen C. Carvedilol in the treatment of chronic heart failure: lessons from the Carvedilol Or Metoprolol European Trial. Vasc Health Risk Manag. 2007;3(1):31-7.

[55] Watanabe K, Arozal W, Sari FR, Arumugam S, Thandavarayan RA, Suzuki K, et al. The Role of Carvedilol in the Treatment of Dilated and Anthracyclines-Induced Cardiomyopathy. Pharmaceuticals. 2011;4(5):7700-781.

[56] Oyama MA, Sisson DD, Prosek R, Bulmer BJ, Luethy MW, Fuentes VL. Carvedilol in dogs with dilated cardiomyopathy. J Vet Intern Med. 2007;21(6):1272-9.

[57] Lyon AR, Sato M, Hajjar RJ, Samulski RJ, Harding SE. Gene therapy: targeting the myocardium. Heart. 2008;94(1):89-99.

[58] Vinge LE, Raake PW, Koch WJ. Gene therapy in heart failure. Circ Res. 2008;102(12):1458-70.

[59] Dieterle T, Meyer M, Gu Y, Belke DD, Swanson E, Iwatate M, et al. Gene transfer of a phospholamban-targeted antibody improves calcium handling and cardiac function in heart failure. Cardiovasc Res. 2005;67(4):678-88.

[60] Paradies P, Carlucci L, Woitek F, Staffieri F, Lacitignola L, Ceci L, et al. Intracoronary Gene Delivery of the Cytoprotective Factor Vascular Endothelial Growth Factor-B167 in Canine Patients with Dilated Cardiomyopathy: A Short-Term Feasibility Study. Vet Sci. 2019;6(1).

[61] Simpson S, Rutland P, Rutland CS. Genomic Insights into Cardiomyo pathies: A Comparative Cross-Species Review. Vet Sci. 2017;4(1).

[62] Kubale V, Prozorowska E, Glocová K, Slater L, Rutland CS. The Function of Seven Transmembrane Receptors in the Cardiovascular System and Their Role in the Development of Cardiomyopathy. Cardiac Diseases. London: IntechOpen; 2020.

[63] Yue Y, Ghosh A, Long C, Bostick B, Smith BF, Kornegay JN, et al. A single intravenous injection of adeno-associated virus serotype-9 leads to whole body skeletal muscle transduction in dogs. Mol Ther. 2008;16(12):1944-52.

[64] Wang Z, Storb R, Halbert CL, Banks GB, Butts TM, Finn EE, et al. Successful regional delivery and long-term expression of a dystrophin gene in canine muscular dystrophy: a preclinical model for human therapies. Mol Ther. 2012;20(8):1501-7.

[65] Sawa Y. Current status of myocardial regeneration therapy. Gen Thorac Cardiovasc Surg. 2013;61(1):17-23.

[66] Borenstein N, Chetboul V, Rajnoch C, Bruneval P, Carpentier A. Successful cellular cardiomyoplasty in canine idiopathic dilated cardiomyopathy. Ann Thorac Surg. 2002;74(1):298-9; author reply 9.

[67] Yoo KJ, Li RK, Weisel RD, Mickle DA, Li G, Yau TM. Autologous smooth muscle cell transplantation improved heart function in dilated cardiomyopathy. Ann Thorac Surg. 2000;70(3):859-65.

[68] Hensley MT, de Andrade J, Keene B, Meurs K, Tang J, Wang Z, et al. Cardiac regenerative potential of cardiosphere-derived cells from adult dog hearts. J Cell Mol Med. 2015;19(8):1805-13.

[69] Hensley MT, Tang J, Woodruff K, Defrancesco T, Tou S, Williams CM, et al. Intracoronary allogeneic cardiosphere-derived stem cells are safe for use in dogs with dilated cardiomyopathy. J Cell Mol Med. 2017;21(8):1503-12.

[70] Cheng Y, Hogarth KA, O'Sullivan ML, Regnier M, Pyle WG. 2-Deoxyadenosine triphosphate restores the contractile function of cardiac myofibril from adult dogs with naturally occurring dilated cardiomyopathy. Am J Physiol Heart Circ Physiol. 2016;310(1):H80-91.

[71] Grillo MP, Markova S, Evanchik M, Trellu M, Moliner P, Brun P, et al. Preclinical in vitro and in vivo pharmacokinetic properties of danicamtiv, a new targeted myosin activator for the treatment of dilated cardiomyopathy. Xenobiotica. 2021;51(2):222-38.

[72] Jung SW, Newhard DK, Harrelson K. Transvenous electrical cardioversion of atrial fibrillation in two dogs. J Vet Cardiol. 2017;19(2):175-81.

[73] Donner J, Kaukonen M, Anderson H, Möller F, Kyöstilä K, Sankari S, et al. Genetic Panel Screening of Nearly 100 Mutations Reveals New Insights into the Breed Distribution of Risk Variants for Canine Hereditary Disorders. PLoS ONE. 2016;11(8): e0161005.

[74] Farrell LL, Schoenebeck JJ, Wiener P, Clements DN, Summers KM. The challenges of pedigree dog health: approaches to combating inherited disease. Canine Genetics and Epidemiology. 2015;2(1):1-14.

[75] The Kennel Club UK. DNA screening schemes and results 2016 [Available from: http://www.thekennelclub.org.uk/health/breeding-for-health/dna-screening-schemes-and-results/.

[76] GTR®. The NIH genetic testing registry 2016 [Available from: http://www.ncbi.nlm.nih.gov/gtr/.

[77] Rubinstein WS, Maglott DR, Lee JM, Kattman BL, Malheiro AJ, Ovetsky M, et al. The NIH genetic testing registry: a new, centralized database of genetic tests to enable access to comprehensive information and improve transparency. Nucleic Acids Research. 2013;41:D925-35.

[78] Lenffer J, Nicholas FW, Castle K, Rao A, Gregory S, Poidinger M, et al. OMIA (Online Mendelian Inheritance in Animals): an enhanced platform and integration into the Entrez search interface at NCBI. Nucleic Acids Research. 2006;34(Database issue):D599-D601.

[79] OMIA. Online Mendelian Inheritance in Animals, OMIA. Faculty of Veterinary Science, University of Sydney 2016 [Available from: http://omia.angis.org.au/.

[80] Tsai KL, Clark LA, Murphy KE. Understanding hereditary diseases using the dog and human as companion model systems. Mammalian Genome. 2007;18(6-7):444-51.

[81] Butler MW, Burt A, Edwards TL, Zuchner S, Scott WK, Martin ER, et al. Vitamin D Receptor Gene as a Candidate Gene for Parkinson Disease. Annals of Human Genetics. 2011;75(2):201-10.

[82] Rivas MA, Beaudoin M, Gardet A, Stevens C, Sharma Y, Zhang CK, et al. Deep resequencing of GWAS loci identifies independent rare variants associated with inflammatory bowel disease. Nature Genetics. 2011;43(11):1066-73.

[83] Saunders CJ, Miller NA, Soden SE, Dinwiddie DL, Noll A, Alnadi NA, et al. Rapid Whole-Genome Sequencing for Genetic Disease Diagnosis in Neonatal Intensive Care Units. Science Translational Medicine. 2012;4(154):154ra35.

[84] Yang Y, Muzny DM, Reid JG, Bainbridge MN, Willis A, Ward PA, et al. Clinical Whole-Exome Sequencing for the Diagnosis of Mendelian Disorders. The New England journal of medicine. 2013;369(16):1502-11.

[85] Mausberg T-B, Wess G, Simak J, Keller L, Drögemüller M, Drögemüller C, et al. A Locus on Chromosome 5 Is Associated with Dilated Cardiomyopathy in Doberman Pinschers. PLoS ONE. 2011;6(5):e20042-e.

[86] Philipp U, Vollmar A, Häggström J, Thomas A, Distl O. Multiple Loci Are Associated with Dilated Cardiomyopathy in Irish Wolfhounds. PLoS ONE. 2012;7(6):e36691-e.

[87] Andrae B, Andersson TML, Lambert PC, Kemetli L, Silfverdal L, Strander B, et al. Screening and cervical cancer cure: population based cohort study. British Medical Journal. 2012;344:e900.

[88] Greenberg B, Butler J, Felker GM, Ponikowski P, Voors AA, Desai AS, et al. Calcium upregulation by percutaneous administration of gene therapy in patients with cardiac disease (CUPID 2): a randomised, multinational, double-blind, placebo-controlled, phase 2b trial. The Lancet. 2016;387(10024):1178-86.

[89] Ho CY, Lakdawala NK, Cirino AL, Lipshultz SE, Sparks E, Abbasi SA, et al. Diltiazem Treatment for Pre-Clinical Hypertrophic Cardiomyopathy Sarcomere Mutation CarriersA Pilot Randomized Trial to Modify Disease Expression. JACC: Heart Failure. 2015;3(2):180-8.

[90] Kaback MM. Population-based genetic screening for reproductive counseling: the Tay-Sachs disease model. Eur J Pediatr. 2000;159(3):S192-S5.

[91] Li X, You R, Wang X, Liu C, Xu Z, Zhou J, et al. Effectiveness of prophylactic surgeries in BRCA1 or BRCA2 mutation carriers: a meta-analysis and systematic review. Clinical Cancer Research. 2016;22(15):3971-81.

[92] Nordestgaard BG, Chapman MJ, Humphries SE, Ginsberg HN, Masana L, Descamps OS, et al. Familial hypercholesterolaemia is

underdiagnosed and undertreated in the general population: guidance for clinicians to prevent coronary heart disease. European Heart Journal. 2013;34(45):3478-90.

[93] Blair SN, Kohl HW, 3rd, Barlow CE, Paffenbarger RS, Jr., Gibbons LW, Macera CA. Changes in physical fitness and all-cause mortality. A prospective study of healthy and unhealthy men. The Journal of the American Medical Association. 1995;273(14):1093-8.

[94] Maeder ML, Gersbach CA. Genome-editing Technologies for Gene and Cell Therapy. Mol Ther. 2016;24(3):430-46.

[95] George R, Kovak K, Cox SL. Aligning Policy to Promote Cascade Genetic Screening for Prevention and Early Diagnosis of Heritable Diseases. Journal of Genetic Counseling. 2015;24(3):388-99.

[96] Ranthe MF, Carstensen L, Øyen N, Jensen MK, Axelsson A, Wohlfahrt J, et al. Risk of Cardiomyopathy in Younger Persons With a Family History of Death from Cardiomyopathy: A Nationwide Family Study in a Cohort of 3.9 Million Persons. Circulation. 2015;132(11):1013-9.

[97] Meurs KM, Lahmers S, Keene BW, White SN, Oyama MA, Mauceli E, et al. A splice site mutation in a gene encoding for PDK4, a mitochondrial protein, is associated with the development of dilated cardiomyopathy in the Doberman pinscher. Human Genetics. 2012;131(8):1319-25.

[98] Meurs KM, Stern JA, Sisson DD, Kittleson MD, Cunningham SM, Ames MK, et al. Association of dilated cardiomyopathy with the striatin mutation genotype in boxer dogs. Journal of veterinary internal medicine / American College of Veterinary Internal Medicine. 2013;27(6):1437-40.

[99] Posafalvi A, Herkert JC, Sinke RJ, van den Berg MP, Mogensen J, Jongbloed JDH, et al. Clinical utility gene card for: dilated cardiomyopathy (CMD). European Journal of Human Genetics. 2012;21(10).

[100] Schatzberg SJ, Olby NJ, Breen M, Anderson LV, Langford CF, Dickens HF, et al. Molecular analysis of a spontaneous dystrophin 'knockout' dog. Neuromuscular Disorders. 1999;9(5):289-95.

[101] Friedenberg SG, Chdid L, Keene B, Sherry B, Motsinger-Reif A, Meurs KM. Use of RNA-seq to identify cardiac genes and gene pathways differentially expressed between dogs with and without dilated cardiomyopathy. American Journal of Veterinary Research. 2016;77(7):693-9.

[102] Simpson S, Edwards J, Emes RD, Cobb MA, Mongan NP, Rutland CS. A predictive model for canine dilated cardiomyopathy-a meta-analysis of Doberman Pinscher data. PeerJ. 2015;3:e842.

[103] Schatzberg SJ, Olby NJ, Breen M, Anderson LV, Langford CF, Dickens HF, et al. Molecular analysis of a spontaneous dystrophin 'knockout' dog. Neuromuscul Disord. 1999;9(5):289-95.

[104] England J, Loughna S, Rutland CS. Multiple Species Comparison of Cardiac Troponin T and Dystrophin: Unravelling the DNA behind Dilated Cardiomyopathy. J Cardiovasc Dev Dis. 2017;4(3).

[105] Biesiadecki BJ, Elder BD, Yu ZB, Jin JP. Cardiac troponin T variants produced by aberrant splicing of multiple exons in animals with high instances of dilated cardiomyopathy. J Biol Chem. 2002;277(52):50275-85.

[106] Wess G, Schulze A, Butz V, Simak J, Killich M, Keller LJM, et al. Prevalence of dilated cardiomyopathy in

Doberman Pinschers in various age groups. Journal of Veterinary Internal Medicine. 2010;24(3):533-8.

[107] Beier P, Reese S, Holler PJ, Simak J, Tater G, Wess G. The role of hypothyroidism in the etiology and progression of dilated cardiomyopathy in Doberman Pinschers. J Vet Intern Med. 2015;29(1):141-9.

[108] Phillips DE, Harkin KR. Hypothyroidism and myocardial failure in two Great Danes. J Am Anim Hosp Assoc. 2003;39(2):133-7.

[109] Calvert CA, Jacobs GJ, Medleau L, Pickus CW, Brown J, McDermott M. Thyroid-stimulating hormone stimulation tests in cardiomyopathic Doberman pinschers: a retrospective study. J Vet Intern Med. 1998;12(5): 343-8.

[110] Tidholm A, Jonsson L. A retrospective study of canine dilated cardiomyopathy (189 cases). J Am Anim Hosp Assoc. 1997;33(6):544-50.

[111] Tidholm A, Haggstrom J, Hansson K. Effects of dilated cardiomyopathy on the renin-angiotensin-aldosterone system, atrial natriuretic peptide activity, and thyroid hormone concentrations in dogs. Am J Vet Res. 2001;62(6):961-7.

[112] Taylor RR, Covell JW, Ross J, Jr. Influence of the thyroid state on left ventricular tension-velocity relations in the intact, sedated dog. J Clin Invest. 1969;48(4):775-84.

[113] Al-Mamun MA, Farid DM, Ravenhil L, Hossain MA, Fall C, Bass R. An in silico model to demonstrate the effects of Maspin on cancer cell dynamics. Journal of Theoretical Biology. 2016;388:37-49.

[114] Betz RC, Petukhova L, Ripke S, Huang H, Menelaou A, Redler S, et al. Genome-wide meta-analysis in alopecia areata resolves HLA associations and reveals two new susceptibility loci. Nature Communications. 2015;6.

[115] Brown DC, Agnello K, Iadarola MJ. Intrathecal resiniferatoxin in a dog model: efficacy in bone cancer pain. Pain. 2015;156(6):1018-24.

[116] Eckstein F, Kwoh CK, Boudreau RM, Wang Z, Hannon MJ, Cotofana S, et al. Quantitative MRI measures of cartilage predict knee replacement: a case–control study from the Osteoarthritis Initiative. Annals of the Rheumatic Diseases. 2013;72(5):707-14.

[117] Lee D-F, Su J, Kim Huen S, Chang B, Papatsenko D, Zhao R, et al. Modeling Familial Cancer with Induced Pluripotent Stem Cells. Cell. 2015;161(2):240-54.

[118] Mariani P, Servois V, De Rycke Y, Bennett SP, Feron JG, Almubarak MM, et al. Liver metastases from breast cancer: Surgical resection or not? A case-matched control study in highly selected patients. European Journal of Surgical Oncology. 2013;39(12): 1377-83.

[119] Simoni RD, Hill RL, Vaughan M. The Discovery of Insulin: the Work of Frederick Banting and Charles Best. Journal of Biological Chemistry. 2002;277(26):e15.

[120] Basso C, Fox PR, Meurs KM, Towbin JA, Spier AW, Calabrese F, et al. Arrhythmogenic Right Ventricular Cardiomyopathy Causing Sudden Cardiac Death in Boxer Dogs: A New Animal Model of Human Disease. Circulation. 2004;109(9):1180-5.

[121] Davidson AG, Bell RJ, Lees GE, Kashtan CE, Davidson GS, Murphy KE. Genetic cause of autosomal recessive hereditary nephropathy in the English Cocker Spaniel. Journal of veterinary internal medicine / American College of

Veterinary Internal Medicine. 2007;21(3):394-401.

[122] Egenvall A, Nødtvedt A, von Euler H. Bone tumors in a population of 400 000 insured Swedish dogs up to 10 y of age: incidence and survival. Canadian Journal of Veterinary Research. 2007;71(4):292-9.

[123] Zangerl B, Goldstein O, Philp AR, Lindauer SJP, Pearce-Kelling SE, Mullins RF, et al. Identical Mutation in a Novel Retinal Gene Causes Progressive Rod-Cone Degeneration (prcd) in Dogs and Retinitis Pigmentosa in Man. Genomics. 2006;88(5):551-63.

[124] Parker HG, Kim LV, Sutter NB, Carlson S, Lorentzen TD, Malek TB, et al. Genetic structure of the purebred domestic dog. Science. 2004;304(5674):1160-4.

[125] Calboli FCF, Sampson J, Fretwell N, Balding DJ. Population Structure and Inbreeding From Pedigree Analysis of Purebred Dogs. Genetics. 2008;179(1):593-601.

[126] Jansson M, Laikre L. Recent breeding history of dog breeds in Sweden: modest rates of inbreeding, extensive loss of genetic diversity and lack of correlation between inbreeding and health. Journal of Animal Breeding and Genetics. 2014;131(2):153-62.

[127] Willet CE, Makara M, Reppas G, Tsoukalas G, Malik R, Haase B, et al. Canine Disorder Mirrors Human Disease: Exonic Deletion in HES7 Causes Autosomal Recessive Spondylocostal Dysostosis in Miniature Schnauzer Dogs. PLoS ONE. 2015;10(2):e0117055.

[128] Marsden CD, Ortega-Del Vecchyo D, O'Brien DP, Taylor JF, Ramirez O, Vilà C, et al. Bottlenecks and selective sweeps during domestication have increased deleterious genetic variation in dogs. Proceedings of the National Academy of Sciences. 2016;113(1):152-7.

[129] Olsson M, Frankowiack M, Tengvall K, Roosje P, Fall T, Ivansson E, et al. The dog as a genetic model for immunoglobulin A (IgA) deficiency: Identification of several breeds with low serum IgA concentrations. Veterinary Immunology and Immunopathology. 2014;160(3-4):255-9.

[130] Decker B, Parker HG, Dhawan D, Kwon EM, Karlins E, Davis BW, et al. Homologous Mutation to Human BRAF V600E Is Common in Naturally Occurring Canine Bladder Cancer—Evidence for a Relevant Model System and Urine-Based Diagnostic Test. Molecular Cancer Research. 2015;13(6):993-1002.

[131] Calderón-Garcidueñas L, Mora-Tiscareño A, Ontiveros E, Gómez-Garza G, Barragán-Mejía G, Broadway J, et al. Air pollution, cognitive deficits and brain abnormalities: A pilot study with children and dogs. Brain and Cognition. 2008;68(2):117-27.

[132] Ostrander EA, Galibert F, Patterson DF. Canine genetics comes of age. Trends in Genetics. 2000;16(3):117-24.

[133] Fosmire SP, Thomas R, Jubala CM, Wojcieszyn JW, Valli VEO, Getzy DM, et al. Inactivation of the p16 Cyclin-Dependent Kinase Inhibitor in High-Grade Canine Non-Hodgkin's T-Cell Lymphoma. Veterinary Pathology. 2007;44(4):467-78.

[134] Lakey JRT, Cavanagh TJ, Zieger MAJ, Wright M. Evaluation of a Purified Enzyme Blend for the Recovery and Function of Canine Pancreatic Islets. Cell Transplantation. 1998;7(4):365-72.

[135] Cooper DS, Doherty GM, Haugen BR, Kloos RT, Lee SL, Mandel SJ, et al. Revised American

Thyroid Association Management Guidelines for Patients with Thyroid Nodules and Differentiated Thyroid Cancer. Thyroid. 2009;19(11):1167-214.

[136] Mant J, Fitzmaurice DA, Hobbs FDR, Jowett S, Murray ET, Holder R, et al. Accuracy of diagnosing atrial fibrillation on electrocardiogram by primary care practitioners and interpretative diagnostic software: analysis of data from screening for atrial fibrillation in the elderly (SAFE) trial. British Medical Journal. 2007;335(7616):380-2.

[137] Porto G, Brissot P, Swinkels DW, Zoller H, Kamarainen O, Patton S, et al. EMQN best practice guidelines for the molecular genetic diagnosis of hereditary hemochromatosis (HH). European Journal of Human Genetics. 2015;24(4):479-95.

[138] Rebbeck TR, Friebel T, Lynch HT, Neuhausen SL, van 't Veer L, Garber JE, et al. Bilateral Prophylactic Mastectomy Reduces Breast Cancer Risk in BRCA1 and BRCA2 Mutation Carriers: The PROSE Study Group. Journal of Clinical Oncology. 2004;22(6):1055-62.

[139] Antoniou A, Pharoah PDP, Narod S, Risch HA, Eyfjord JE, Hopper JL, et al. Average Risks of Breast and Ovarian Cancer Associated with BRCA1 or BRCA2 Mutations Detected in Case Series Unselected for Family History: A Combined Analysis of 22 Studies. The American Journal of Human Genetics. 2003;72(5):1117-30.

Section 2

One Health

Chapter 3

The State of Knowledge on Intestinal Helminths in Free-Roaming Dogs in Southern South America

Luciano Ritossa, Gustavo Viozzi and Verónica Flores

Abstract

In South America there are more dogs per person than in developed countries. Many owners allow their dogs to roam freely in public areas, which favours the spread of zoonotic diseases. The objective of this work is to describe, through bibliographic analysis, the occurrence, prevalence, species richness, and distribution of intestinal helminth parasites found in dog faeces from urban and rural areas of southern South America (Argentina-Chile-Uruguay). Using three databases, we performed a systematic review of articles published between 2000 and 2020 in indexed journals. A total of 219 articles was evaluated for eligibility, and of these 67 were included in the final analysis; 48 correspond to Argentina, 17 to Chile, and 2 to Uruguay. The total number of parasite taxa recorded was 22, the most frequently occurring species being *Toxocara canis*, *Ancylostoma* sp., *Trichuris vulpis* and *Echinococcus* sp. Species richness was correlated with sample size and varied between 1 and 10 species. In addition, disease risk is not homogeneously distributed. Due to the high infection levels in dogs, urban and rural dwellers are at risk of infection with zoonotic diseases transmitted by these animals, therefore a One Health approach to public health would be advisable.

Keywords: Argentina, Chile, Uruguay, Helminths, Canine faeces, *Toxocara canis*, *Echinococcus granulosus*, *Ancylostoma caninum*, *Trichuris vulpis*, systematic bibliographic review, zoonotic risk

1. Introduction

1.1 Dog populations

Humans and dogs share a long history and were probably associated with European early-modern humans [1], coexisting indoors and outdoors and colonising new environments, often in cooperation [2]. From ancient times dogs have been used by humans as tools for different purposes, such as hunting, gathering food, caring for livestock, protection, and more recently as detectors of explosives and drugs, as companion animals, or as assistants for people with various types of disease or disability [3–5]. Therefore, their coexistence has been wide-ranging, and has generated numerous opportunities for around 260 zoonotic diseases to emerge between dogs and humans [2, 6].

There are almost one billion dogs worldwide [7], but the relationship between the numbers of people and dogs varies according to the geographic area and socioeconomic conditions of each country or region [8]. In developed countries the human to dog ratio varies from 6 to 10:1 according to the World Health Organisation [9]; in Italy the human:dog ratio is 9:1 [10], and in the United States it is 3.6:1 [11]. The dog population in South America is very large, around 87.6 million. In Brazil in particular there are 44.9 million children aged under 14 years, and an estimated total of 52.2 million dogs, which means there are more dogs than children [12]. In Argentina, a survey carried out for food companies determined that there are approximately 9 million dogs, and that 78% of households have a dog, whose function is mainly exclusively companionship [13]. The situation in Chile is similar, where the dog population is around 3.5 million and 64% of households have at least one [14], while in Uruguay the dog population is 1.75 million and 72% of households own a dog [15].

To encourage responsible ownership of this large number of dogs, it was necessary to enact laws indicating what responsible dog care implies (Argentina: Decree 1088/11; Chile: No. 21.020/17; Uruguay: No. 1189/14). Animal welfare thus imposes obligations on the owner, which include vaccinations, deworming, neutering, adequate food, and keeping pets confined to the household or taking them outside on a lead, thus preventing them from roaming freely. It should be noted that in most localities of these countries these laws are not enforced effectively [16].

1.2 Dog care

Although national laws have been promulgated several years ago, knowledge of them and the care received by dogs is far from adequate [17–20]. The biggest problem in these countries is that dogs are allowed to roam freely in public areas, and this is associated with education, socio-economic level, the idiosyncrasy and customs of each country, the role the dog plays within the family, and the low importance that people give to how their dog can affect other people or animals [21]. In addition, allowing dogs to roam freely is strongly correlated with other aspects of dog care, such as a lack of appropriate vaccination and deworming treatment [21]. The care given to dogs that roam freely is poorer than for dogs which are confined, and they are rarely taken to the vet due to the high cost that this represents [22]. In Chile, the average cost spent per pet for annual veterinary check-ups, diagnoses, vaccines and treatment is US$ 330 [4], while in Argentina this cost is around US$ 100 annually (personal observation). The percentage of vaccinated dogs is low, even when there is a possibility of rabies contagion [14, 23], and the frequency of deworming is in most cases inadequate considering that dogs can roam freely on public roads, becoming reinfected [23–25]. The percentage of animals that are neutered is also insufficient, despite the national or local neutering programs run in the three countries [21, 26, 27]. Neutered animals represent less than half the dog population [21, 23, 28] and the majority are older than 3 years; in many cases dogs are allowed to have at least one litter of offspring [23].

1.3 Dogs, parasites and diseases

One Health is recognised as a valuable paradigm for global health management, and seeks the integration of human and animal health. The risk of transmission of a zoonotic disease from dogs to humans is related to the abundance of infectious forms in the environment, climatic conditions, whether dogs roam freely, and the behaviour of humans that exposes them to infective sources [29, 30]. It has been observed that free-roaming dogs are more exposed and prone to acquiring parasites [24, 31–33]. In Chile, rural dogs are associated with agricultural and livestock activities. They are unsupervised, have freedom to roam and are given limited veterinary care [34]. In Argentina, parasite richness and prevalence are positively associated with free-roaming animals, and only a small proportion of dogs (17%) is subjected to some degree of movement restriction [20]. In the cities of Argentinian Patagonia, another important factor that promotes infection by zoonotic parasites, mainly cystic echinococcosis, is the domestic slaughter of small ruminants for human consumption. This practice occurs frequently in rural areas and the peripheral low-income neighbourhoods of cities, where dogs are fed with the raw offal of sheep and goats [35, 36]. The vast majority of parasites registered in South America are cosmopolitan zoonotic parasites transmitted through dog faeces, such as *Toxocara canis*, *Ancylostoma caninum*, *Toxascaris leonina*, *Echinococcus* spp., and *Dipylidium caninum*, which are common parasites in dogs worldwide [12]. Zoonotic parasitic infections in dogs are a public health issue not only in developing countries but also in developed nations, such as in the USA and European countries [37, 38]. Other parasites like *Trichuris vulpis* are distributed worldwide, but are rarely transmitted to humans [39]. Some human parasites like *Ascaris lumbricoides* and *Strongyloides stercoralis* are occasionally reported in dogs [40, 41]. Therefore, worldwide, dogs may harbour zoonotic parasites that affect the health and wellbeing of humans, their distribution being linked to poverty, poor knowledge of sanitary practices, insufficient hygiene and problems with unconfined and untreated dogs [42]. Pet diseases may pose risks to human health but are rarely included in surveillance systems. Although pet-borne infections have become increasingly relevant to human health, systematic notification of these infections is not currently conducted, except for rabies and Echinococcosis in some countries [22, 43].

Southern South America is a region with varied geography and climate and marked altitudinal and latitudinal differences; for example, plains (Pampas in Argentina and Uruguay), arid plateaus (Patagonia), forests (Patagonia and northeastern Argentina), and mountains of high altitude between Argentina and Chile (the Andes). The climate ranges from humid tropical in northern Argentina and Uruguay, arid in northern Chile, to humid cold in the south of Argentina and Chile. This climatic variety favours the distribution and occurrence of different parasites. On the other hand, the socio-economic condition of a large part of the population is characterised by poverty and a low-income economy. This scenario is accompanied by a lack of parasitological studies, surveillance and zoonosis control plans on the part of public health organisations [44].

The objective of this work is to describe, through bibliographic analysis, the occurrence, prevalence, species richness, and distribution of intestinal helminth parasites found in dog faeces in urban and rural areas of southern South America (Argentina-Chile-Uruguay).

2. Materials and methods

2.1 Search approach

Three databases (PubMed, Google Scholar and Scopus) were searched for studies published between 2000 and 2020. The search terms were “dog AND parasite AND Argentina”; “dog AND parasite AND Chile”; and “dog AND parasite AND Uruguay”.

The Google Scholar search in particular returned a large number of results, of which the first 700 titles were read (and in some cases the abstract); however, it was observed that after the first 200 no results were found that met the search requirements.

2.2 Paper assortment

The studies to be included were identified independently by two reviewers, and were confirmed by a third reviewer following standardised methodology [45]. The studies included met the following criteria: (1) full text articles available online; (2) published between 2000 and 2020; (3) peer-reviewed, original papers published either in English or Spanish; (4) cross-sectional studies that assessed the prevalence of any intestinal helminth parasite of dogs in Argentina, Chile or Uruguay; (5) studies that detected parasite infection in faeces using at least one parasitological, serological and/or molecular method; (6) studies that reported sample sizes, and the prevalence of each parasite species. Reviews and case reports were excluded. The following data were extracted from each article: authors, publication year, country, localities (coordinates), type of locality (rural/urban), sample size, detection method, prevalence of each parasite, number of parasite species.

2.3 Parasite distribution

The distribution maps were constructed using the Free and Open Source Geographic Information System (QGis system). The coordinates for the site locations were taken from the selected works or were completed using Google Earth. The prevalence values shown on the maps were obtained from the studies included in the bibliographic review. The map of South America was obtained from shape files from *Instituto Geográfico Nacional* [46].

2.4 Statistical analysis

Spearman's rank Correlation Tests were performed to analyse the relation between richness, with sample size and latitude. All sites with richness = 1 were excluded, since they searched for only one parasite.

3. Results

From the search in the 3 databases, 29,450 scientific items were found. Of these, 24,517 belong to the period between 2000 and 2020. After analysing the titles and abstracts, 24,298 articles were excluded because they did not comply with the objectives or inclusion criteria, did not include helminths, did not correspond to the countries under study, or were not cross-sectional studies. A total of 219 articles were evaluated for eligibility. After removing the duplicates, 67 were included in the final analysis (**Table 1**), and the full texts of these relevant articles were reviewed in depth. Forty-eight corresponded to Argentina, 17 to Chile, and 2 to Uruguay (**Figure 1**). The data come from analysis of 32,300 dog faeces collected in urban or rural sites of the 3 countries. Sample sizes in the different studies ranged from 4 to 2,417, except for Uruguay where 5,356 faeces were analysed for the National Echinococcosis Control Programs, without considering the presence of other parasites (**Table 1**).

The number of copro-parasitological techniques used in each study varied between 1 and 3, with a total of 15 different methods (**Table 1**). The most

Autor	Año	Country	Name Study Locality	Coordinates	Sample size	Fixing method	of tection Methods	No. Of detection methods	RURAL	URBAN	Richness	Ancylostomids	*Ancylostoma* sp.	*Uncinaria* sp.	*Ascaris* sp.	*Dibothriocephalus* sp.	*Dipylidium caninum*	*Echinococcus* sp.	*Eucoleus aerophila*	*Eucoleus boehmi*	Capillaria sp.	*Taenia multiceps* sp.	*Strongyloides* eggs	*Spirocerca*	*Taenidae*	*Taenia hydatigena*	*Taenia ovis*	*Toxascaris* sp.	*Toxocara* sp.	*Trichuris* sp.	Trematodes	*Oncicola canis*	*Physalopetra* sp.
Acosta Jamett et al. [47]	2010	Chile	Tangue	30°20'S, 71°34'W	120		ELISA	1	rural		1							10															
Acosta Jamett et al. [47]	2010	Chile	Guanaqueros	30°11'S, 71°25'W	81		ELISA	1		urban	0																						
Acosta Jamett et al. [47]	2010	Chile	Coquimbo	29°57'S, 71°20'W	128		ELISA	1		urban	1							15															
Acosta Jamett et al. [48]	2014	Chile	Combarbalá	31°10'S, 71°03'W	52		CoproElisa	1		urban	1							27															
Andresiuk et al. [49]	2007	Argentina	Mar del Plata	37°56'S, 57°35'W	400		Willis Flotation	1		urban	3		62.8																13.9	46.75			
Andresiuk et al. [50]	2003	Argentina	Mar del Plata	38°00'S, 57°33'W	125		Flotation, sedimentation of Willis	1		urban	4		62.96	24.07							2.56								22.22	62.96			
Andresiuk et al. [29]	2004	Argentina	Mar del Plata	38°00'S, 57°33'W	288		Flotation, sedimentation of Willis	1		urban	3			65.83															14.17	46.67			
Archelli et al. [51]	2018	Argentina	Ensenada	34°51'S, 57°54'W	217	Formol 10%	Sedimentation of Teleman and Flotation of Sheater	2		urban	1																		23.0				
Arezo et al . [36]	2020	Argentina	Bariloche	41°10'S, 71°18'W	1780		coproElisa Echinococcus	1	rural		1							1															
Arezo et al . [36]	2020	Argentina	El Bolson	41°58'S, 71°32'W			CoproElisa	1	rural																								
Arezo et al. [36]	2020	Argentina	Comallo	41°02'S, 70°16'W			CoproElisa	1	rural		1							1															
Arezo et al. [36]	2020	Argentina	El Cuy	39°56'S, 68°20'W			CoproElisa	1	rural		1							1															
Arezo et al. [36]	2020	Argentina	Ing. Jacobacci	41°18'S, 69°35'W			CoproElisa	1	rural		1							1															
Arezo et al. [36]	2020	Argentina	Maquinchao	41°15'S, 68°42'W			CoproElisa	1	rural																								
Arezo et al. [36]	2020	Argentina	Los Menucos	40°50'S, 68°05'W			CoproElisa	1	rural		1							1															
Arezo et al. [36]	2020	Argentina	Ñorquinco	41°51'S, 70°54'W			CoproElisa	1	rural		1							1															
Arezo et al. [36]	2020	Argentina	Pilcaniyeu	41°07'S, 70°43'W			CoproElisa	1	rural		1							1															

Autor	Año	Country	Name Study Locality	Coordinates	Sample size	Fixing method	of tection Methods	No. Of detection methods	RURAL	URBAN	Richness	Ancylostomids	*Ancylostoma* sp.	*Uncinaria* sp.	*Ascaris* sp.	*Dibothriocephalus* sp.	*Dipylidium caninum*	*Echinococcus* sp.	*Eucoleus aerophila*	*Eucoleus boehmi*	Capillaria sp.	*Taenia multiceps* sp.	*Strongyloides* eggs	*Spirocerca*	*Taenidae*	*Taenia hydatigena*	*Taenia ovis*	*Toxascaris* sp.	*Toxocara* sp.	*Trichuris* sp.	Trematodes	*Oncicola canis*	*Physalopetra* sp.
Arezo et al. [36]	2020	Argentina	Ramos Mexia	40°30′S, 67°17′W			CoproElisa	1	rural		1							1															
Arezo et al. [36]	2020	Argentina	Sierra Colorada	40°35′S, 67°45′W			CoproElisa	1	rural		1							1															
Arezo et al. [36]	2020	Argentina	Sierra Grande	41°36′S, 65°21′W			CoproElisa	1	rural																								
Arezo et al . [36]	2020	Argentina	Valcheta	40°42′S, 66°09′W			CoproElisa	1	rural		1							1															
Armstrong et al. [52]	2011	Chile	Temuco	37°24′S, 72°31′W	196		Flotation with zinc	1		urban	4												9.3			4.7			12.4	4.7			
Casas et al. [53]	2013	Argentina	La Quiaca	22°06′S, 65°36′W	89		Copro, Elisa and WB	2		urban	1							2.2															
Castillo et al. [54]	2000	Chile	Santiago de Chile	33°27′S, 70° 40′W	288	Formol salino	Telemann modified, using ethandl acetate	1		urban	4		4.5		0.7														13.5	7.3			
Chiodo et al. [55]	2006	Argentina	General Mansilla	35°04′S, 57°44′W	81		Sedimentation of Teleman modified	1	rural		1																		6.17				
Cociancic et al. [56]	2017	Argentina	La Plata	34°56′S, 57°57′W	78		Sedimentation of Ritchie and Flotation of Willis	2		urban	7		69.2	41.0			1.3				1.3							1.3	21.8	28.2			
Cociancic et al. [32]	2020	Argentina	Ushuaia	54°48′S, 68°18′W	80	Formol 5%	Sedimentation and Flotac	2		urban	7			1.3				2.5											5.0	1.3			
De Costas et al. [57]	2014	Argentina	Tumbaya	23°51′S, 65°28′W	222		Copro, Elisa and WB	2			1							11.7															
De Costas et al. [57]	2014	Argentina	Humahuaca	23°12′S, 65°21′W	18		Copro, Elisa and WB	2			1							27.7															
De Costas et al. [57]	2014	Argentina	Tilcara	23°34′S, 65°23′W	64		Copro, Elisa and WB	2			1							14.0															
De Costas et al. [57]	2014	Argentina	Cochinoca	22°44′S, 65°53′W	94		Copro, Elisa and WB	2			1							9.5															
De Costas et al. [57]	2014	Argentina	Susques	23°24′S, 66°22′W	50		Copro, Elisa and WB	2			1							2.0															
De Costas et al. [57]	2014	Argentina	Santa Catalina	21°56′S, 66°03′W	28		Copro, Elisa and WB	2			1							10.7															
De Costas et al. [57]	2014	Argentina	Yavi	22°07′S, 65°27′W	47		Copro, Elisa and WB	2			1							14.8															

Autor	Año	Country	Name Study Locality	Coordinates	Sample size	Fixing method	of tection Methods	No. Of detection methods	RURAL	URBAN	Richness	Ancylostomids	*Ancylostoma* sp.	*Uncinaria* sp.	*Ascaris* sp.	*Dibothriocephalus* sp.	*Dipylidium caninum*	*Echinococcus* sp.	*Eucoleus aerophila*	*Eucoleus boehmi*	Capillaria sp.	*Taenia multiceps* sp.	*Strongyloides* eggs	*Spirocerca*	*Taenidae*	*Taenia hydatigena*	*Taenia ovis*	*Toxascaris* sp.	*Toxocara* sp.	*Trichuris* sp.	Trematodes	*Oncicola canis*	*Physalopetra* sp.
Dopchiz et al. [58]	2013	Argentina	Lobos, Bs As	35°10′S, 59°05′W	42	Formol 10%, freezado	Sedimetation of Ritchie, Flotation of Sheater and CoproElisa	3	rural		6		11.9	14.29				19.05	26.19											26.19			
Enriquez et al. [59]	2019	Argentina	Pampa del Indio, Chaco	26°02′S, 59°55′W	85	SAF solution	Flotation with NaCl and Sedimentation	2		urban	8		68.2				2.4				1.2			5.9	5.9				14.1	3.5	15.3		
Flores et al. [35]	2017	Argentina	Bariloche	41°10′S, 71°18′W	118		Sheater Flotation	1		urban	9		47.0			16.9	0.8	9.3			5.1		2.5					11.9	12.7	39.0			
Fontanarosa et al. [60]	2006	Argentina	Lanus	34°22′S, 58°22′W	262		Sheater Flotation	1		urban	5		9.1															0.05	12.6	11			
Fontanarosa et al. [60]	2006	Argentina	Avellaneda	34°39′S, 58°22′W	547		Sheater Flotation	1		urban	5		8.9				0.8												14.2	5.4			
Fontanarosa et al. [60]	2006	Argentina	Alte Brown	34°50′S, 58°23′W	458		Sheater Flotation	1		urban	5		19																8.9	14.1			
Fontanarosa et al. [60]	2006	Argentina	E.Echeverria	34°52′S, 58°28′W	134		Sheater Flotation	1		urban	5		21.6																6.7	17.9			
Fontanarosa et al. [60]	2006	Argentina	Lomas de Zamora	34°45′S, 58°25′W	499		Sheater Flotation	1		urban	5		13																9.8	10.2			
Fontanarosa et al. [60]	2006	Argentina	Quilmes	34°15′S, 58°15′W	293		Sheater Flotation	1		urban	5		13.6																10.2	7.5			
Gamboa et al. [61]	2011	Argentina	La Plata	34°56′S, 57°53′W	12	Formol 10%	Sedimentation of Ritchie and Flotation of Willis	2		urban	4		16		16														16	8			
Gamboa et al. [62]	2009	Argentina	La Plata Norte	34°56′S, 57°57′W	5		Sedimentation of Ritchie and Carles Barthelemand, and Flotation of Fülleborn	3		urban	4		16.7		16.7														16.7	8.3			
Gamboa et al. [62]	2009	Argentina	La Plata Sur	34°56′S, 57°57′W	4		Sedimentation of Ritchie and Carles Barthelemand, and Flotation of Fülleborn	3		urban	2		33.3																8.3				
Gamboa et al. [62]	2009	Argentina	Aristóbulo del Valle	27°05′S, 54°53′W	11		Sedimentation of Ritchie and Carles Barthelemand, and Flotation of Fülleborn	3		urban	4		90.9		9.1														27.3	9.1			
Gonzalez Acuña et al. [63]	2008	Chile	Archipiélago de Juan Fernández	33°38′S, 78°50′w	40	SAF solution	Teuscher Methods or Flotation of Willis	2	rural		3	30.0						3.9					15										
Gorman et al. [31]	2006	Chile	Santiago de Chile	33°27′S, 70° 40′W	582		Flotation zinc sulfate and Sedimentation of Teleman modified	2		urban	5	5.3					2.1											2.4	9.1	8.6			
Irabedra et al. [64]	2016	Uruguay					5356		CoproElisa	1							3.6																

Autor	Año	Country	Name Study Locality	Coordinates	Sample size	Fixing method	of tection Methods	No. Of detection methods	RURAL	URBAN	Richness	Ancylostomids	*Ancylostoma* sp.	*Uncinaria* sp.	*Ascaris* sp.	*Dibothriocephalus* sp.	*Dipylidium caninum*	*Echinococcus* sp.	*Eucoleus aerophila*	*Eucoleus boehmi*	Capillaria sp.	*Taenia multiceps* sp.	*Strongyloides* eggs	*Spirocerca*	*Taenidae*	*Taenia hydatigena*	*Taenia ovis*	*Toxascaris* sp.	*Toxocara* sp.	*Trichuris* sp.	Trematodes	*Oncicola canis*	*Physalopetra* sp.
Irabedra et al. [64]	2016	Uruguay					1496		CoproElisa		1							7.35															
La Sala et al. [65]	2015a	Argentina	Bahía Blanca	38°44′S, 62°16′W	475	Formol 10%	Sedimentation of Ritchie	1		urban	5		21.1												0.6				2.3	18.1			
La Sala et al. [66]	2015b	Argentina	Bahia Blanca	38°43′S, 62°16′W	475		Direct observation	1		urban	5		22.3												0.6				2.3	18.1			
Lamberti et al . [67]	2014	Argentina	Gra. Pico	35°39′S, 63°45′W	785		Flotation with ClNa	1		urban	3		45.4		7.1															25.8			
Lamberti et al. [68]	2015	Argentina	Gral Pico	35°40′S, 63°44′W	1229		Flotation with ClNa and ZnSO4	2		urban	3		45.4		6.4															21.9			
Larrieu et al. [69]	2014	Argentina	El Bolsón	41°58′S, 71°32′W	68		Copro, Elisa and WB	2	rural		1							11.8															
Larrieu et al. [69]	2014	Argentina	El Cuy	39°56′S, 68°20′W	81		Copro, Elisa and WB	2	rural		1							6.1															
Larrieu et al. [69]	2014	Argentina	Ñorquinco	41°51′S, 70°54′W	47		Copro, Elisa and WB	2	rural		1							6.4															
Larrieu et al. [69]	2014	Argentina	Pilcaniyeu	41°07′S, 70°43′W	19		Copro, Elisa and WB	2	rural		1							5.3															
Larrieu et al. [69]	2014	Argentina	Comallo	41°02′S, 70°16′W	12		Copro, Elisa and WB	2	rural		1							8.3															
Larrieu et al. [69]	2014	Argentina	Ingeniero Jacobacci	41°18′S, 69°35′W	108		Copro, Elisa and WB	2	rural		1							7.4															
Larrieu et al. [69]	2014	Argentina	Maquinchao	41°15′S, 68°42′W	16		Copro, Elisa and WB	2	rural		1							12.5															
Larrieu et al. [69]	2014	Argentina	Los Menucos	40°50′S, 68°05′W	37		Copro, Elisa and WB	2	rural		1							5.4															
Larrieu et al. [69]	2014	Argentina	Sierra Colorada	40°35′S, 67°45′W	42		Copro, Elisa and WB	2	rural		1							2.4															
Larrieu et al. [69]	2014	Argentina	Valcheta	40°42′S, 66°09′W	106		Copro, Elisa and WB	2	rural		1							4.7															
Larrieu et al. [69]	2014	Argentina	Sierra Grande	41°36′S, 65°21′W	14		Copro, Elisa and WB	2	rural		1							7.2															
Lavallén et al. [70]	2011	Argentina	Gral Pueyrredon	38°00′S, 57°33′W	46	Formol 10%	Sediemtation of Ritchie and Flotation of Sheater and coproELISA	3		urban	6		71.74	41.3				8.6	17.36										63.04	45.65			

Autor	Año	Country	Name Study Locality	Coordinates	Sample size	Fixing method	of tection Methods	No. Of detection methods	RURAL	URBAN	Richness	Ancylostomids	*Ancylostoma sp.*	*Uncinaria sp.*	*Ascaris sp.*	*Dibothriocephalus sp.*	*Dipylidium caninum*	*Echinococcus sp.*	*Eucoleus aerophila*	*Eucoleus boehmi*	Capillaria sp.	*Taenia multiceps sp.*	*Strongyloides eggs*	*Spirocerca*	*Taenidae*	*Taenia hydatigena*	*Taenia ovis*	*Toxascaris sp.*	*Toxocara sp.*	*Trichuris sp.*	Trematodes	*Oncicola canis*	*Physalopetra sp.*
Lopez et al. [71]	2006	Chile	Santiago de Chile	33°27′S, 70° 40′W	972	PAF fenol, alcohol and formaldehído	Burrows Technique	1		urban	7	1.8					2.2									0.4		1.4	11.1	8.9			1.2
Luzio et al. [72]	2013	Chile	Tomé	36°37′S, 72°57′W	223	PAF fenol, alcohol and formaldehído	Burrows Technique	1		urban	9		8.1	0.9			9.9				2.7		3.1			1.8		6.3	22.9	8.1			
Luzio et al. [73]	2015	Chile	Santa de los Angeles	37°28′S, 72°21′W	452	PAF fenol, alcohol and formaldehído	Burrows Technique	2		urban	7		4.2	0.44			2.6						0.44			1.6		1.3	9.3				
Luzio et al. [74]	2017	Chile	Concepcion	36°49′S, 73°03′W	64	PAF fenol, alcohol and formaldehído	Burrows Technique	1		urban	5		8.5					29					4.5			6.3			29.7				
Madrid et al. [75]	2008	Argentina	Mar del Plata	38°00′S, 57°33′W	358		Flotation with NaCl	1		urban	7		18.9	11.5							1.1							0.6	5.9	13.4			
Marder et al. [76]	2004	Argentina	Ciudad de Corrientes	27°25′S, 58°52′W	900		Flotation of Willis, Sheater and Faust	3		urban	3		64.5																7.6	3.1			
Martin et al. [77]	2008	Argentina	Paraná	31°44′S, 60°31′W	61	Solución salina 5%	Concentration methods	1		urban	2		67.0																7.0				
Martin et al. [77]	2008	Argentina	Santa Fé	31°38′S, 60°42′W	200	Solución salina 5%	Concentration methods	1		urban	3		14.0																62.0	12.0			
Martin et al. [77]	2008	Argentina	Avellaneda (Santa Fé)	29°07′S, 59°39′W	15	Solución salina 5%	Concentration methods	1		urban	3		5.0																6.0	1.0			
Martin et al. [77]	2008	Argentina	Reconquista (Santa Fé)	29°09′S, 59°39′W	10	Solución salina 5%	Concentration methods	1		urban	2		5.0																5.0				
Martin et al. [77]	2008	Argentina	Calchaquí (Santa Fé)	29°53′S, 60°16′W	17	Solución salina 5%	Concentration methods	1		urban	3		2.0																5.0	1.0			
Martin et al. [77]	2008	Argentina	Hersilia (Santa Fé)	30°00′S, 61°51′W	12	Solución salina 5%	Concentration methods	1		urban	3		4.0																5.0	1.0			
Martin et al. [77]	2008	Argentina	San Carlos Centro (Santa Fé)	31°44′S, 61°06′W	24	Solución salina 5%	Concentration methods	1		urban	3		8.0																6.0	3.0			
Martin et al. [77]	2008	Argentina	Santo Tomé (Santa Fé)	31°40′S, 60° 46′W	54	Solución salina 5%	Concentration methods	1		urban	3		9.0																5.0	2.0			
Mercado et al. [78]	2004	Chile	Arica	18°28′S, 70°19′W	50		Sedimentation and Harada, Mori	2		urban	2		2																4				
Mercado et al. [78]	2004	Chile	Antofagasta	23°38′S, 70°23′W	50		Sedimentation and Harada, Mori	2		urban	2																		2				

Autor	Año	Country	Name Study Locality	Coordinates	Sample size	Fixing method	of tection Methods	No. Of detection methods	RURAL	URBAN	Richness	Ancylostomids	*Ancylostoma* sp.	*Uncinaria* sp.	*Ascaris* sp.	*Dibothriocephalus* sp.	*Dipylidium caninum*	*Echinococcus* sp.	*Eucoleus aerophila*	*Eucoleus boehmi*	*Capillaria* sp.	*Taenia multiceps* sp.	*Strongyloides* eggs	*Spirocerca*	*Taenidae*	*Taenia hydatigena*	*Taenia ovis*	*Toxascaris* sp.	*Toxocara* sp.	*Trichuris* sp.	Trematodes	*Oncicola canis*	*Physalopetra* sp.
Mercado et al. [78]	2004	Chile	Illapel	31°37′S, 71°10′W	50		Sedimentation and Harada, Mori	2		urban	2		7.2																10				
Mercado et al. [78]	2004	Chile	Viña del Mar	33°01′S, 71°33′W	27		Sedimentation and Harada, Mori	2		urban	2																						
Mercado et al. [78]	2004	Chile	Valparaiso	33°02′S, 71°37′W	40		Sedimentation and Harada, Mori	2		urban	2		10																12.5				
Mercado et al. [78]	2004	Chile	San Felipe	32°45′S, 70°43′W	44		Sedimentation and Harada, Mori	2		urban	2																		6.8				
Mercado et al. [78]	2004	Chile	Santiago de Chile	33°27′S, 70° 40′W	54		Sedimentation and Harada, Mori	2		urban	2																		1.9				
Mercado et al. [78]	2004	Chile	Rancagua	34°09′S, 70°44′W	27		Sedimentation and Harada, Mori	2		urban	2																		7.4				
Mercado et al. [78]	2004	Chile	San Fernando	34°35′S, 70°59′W	50		Sedimentation and Harada, Mori	2		urban	2		24																8				
Mercado et al. [78]	2004	Chile	Concepcion	36°49′S, 73°03′W	49		Sedimentation and Harada, Mori	2		urban	2		8.2																6.1				
Mercado et al. [78]	2004	Chile	Temuco	38°44′S, 72°35′W	50		Sedimentation and Harada, Mori	2		urban	2		40																4				
Mercado et al. [78]	2004	Chile	Valdivia	39°48′S, 73°14′W	50		Sedimentation and Harada, Mori	2		urban	2		20																4				
Mercado et al. [78]	2004	Chile	Punta Arenas	53°09′S, 70°54′W	54		Sedimentation and Harada, Mori	2		urban	2																		1.9				
Milano et al. [79]	2005	Argentina	Ciudad de Corrientes	27°25′S, 58°52′W	61	Formol 10%	Sedimentation and flotation of Willis	2		urban	4		32.8				1.6												26.3	3.3			
Milano et al. [79]	2005	Argentina	Ciudad de Corrientes	27°25′S, 58°52′W	40	Formol 10%	Sedimentation and flotation of Willis	2		urban	4		35.0			2.5													17.5	2.5			
Milano et al. [79]	2005	Argentina	Ciudad de Corrientes	27°25′S, 58°52′W	40	Formol 10%	Sedimentation and flotation of Willis	2		urban	4		35.0			12.5													12.5	10.0			
Milano et al. [79]	2005	Argentina	Ciudad de Corrientes	27°25′S, 58°52′W	32	Formol 10%	Sedimentation and flotation of Willis	2		urban	3		50.0																15.6	9.4			
Milano et al. [79]	2005	Argentina	Ciudad de Corrientes	27°25′S, 58°52′W	40	Formol 10%	Sedimentation and flotation of Willis	2		urban	4		45.0			2.5													20.0	2.5			
Milano et al. [79]	2005	Argentina	Ciudad de Corrientes	27°25′S, 58°52′W	33	Formol 10%	Sedimentation and flotation of Willis	2		urban	3		48.2																6.0	3.0			

Autor	Año	Country	Name Study Locality	Coordinates	Sample size	Fixing method	of tection Methods	No. Of detection methods	RURAL	URBAN	Richness	Ancylostomids	*Ancylostoma* sp.	*Uncinaria* sp.	*Ascaris* sp.	*Dibothriocephalus* sp.	*Dipylidium caninum*	*Echinococcus* sp.	*Eucoleus aerophila*	*Eucoleus boehmi*	*Capillaria* sp.	*Taenia multiceps* sp.	*Strongyloides* eggs	*Spirocerca*	*Taenidae*	*Taenia hydatigena*	*Taenia ovis*	*Toxascaris* sp.	*Toxocara* sp.	*Trichuris* sp.	Trematodes	*Oncicola canis*	*Physalopetra* sp.
Milano et al. [79]	2005	Argentina	Ciudad de Corrientes	27°25′S, 58°52′W	34	Formol 10%	Sedimentation and flotation of Willis	2		urban	3		38.2																17.6	5.9			
Milano et al. [79]	2005	Argentina	Ciudad de Corrientes	27°25′S, 58°52′W	44	Formol 10%	Sedimentation and flotation of Willis	2		urban	4		43.2			4.5													6.8				
Milano et al. [79]	2005	Argentina	Ciudad de Corrientes	27°25′S, 58°52′W	38	Formol 10%	Sedimentation and flotation of Willis	2		urban	4		50.0			2.6													15.8	7.9			
Motta et al . [80]	2019	Argentina	Rio Cuarto	33°07′S, 64°20′W	493	Formol 10%	Flotation of Willis, and Sheather, and Sedimentation	3		urban	5		30.83				0.61			1.42									6.9	9.94			
Natalini et al . [81]	2020	Argentina	Parque Nac Mburucuya	27°58′S, 57°59′W	28	Formol 10%	Flotation Sheater and sedimentation of Ritchie	2	rural		6			4	4		4		7						14				4	4			
Natalini et al . [81]	2020	Argentina	San Nicolás NP	27°59′S, 57°35′W	23	Formol 10%	Flotation Sheater and Sedimentation of Ritchie	2	rural		3		52						9				9										
Oku et al. [82]	2004	Uruguay	Tacuarembo	31°42′S, 55°58′W	79		Necropsy	1		urban	4						38	23				1				8							
Oku et al. [82]	2004	Uruguay		31°45′S, 55°58′W	- 31.75		Necropsy	1	rural		6					1	30	3.49				4				23	3						
Olivares et al. [83]	2014	Chile	Temuco	37°24′S, 72°36′W	102		Flotation and Sedimentation of Teuscher	1		urban	4										21.5		12.7						21.5	35.2			
Opazo et al. [84]	2019	Chile	Valparaiso	33°02′S, 71°37′W	30	PAF fenol, alcohol and formaldehído	Burrows Technique	1	rural		6	7		13			17						17						40	3			
Oyarzun et al. [85]	2019	Chile	Contulmo	38°00′S, 73°14′W	270	Alcohol	Sedimentation and Flotation of Teuscher	1	rural		5			25.5				4					53.9			30.5				15.6			
Parra et al. [86]	2017	Argentina	Ancajuli	26°35′S, 65°33′W	43		CoproElisa	1	rural		1							13															
Parra et al. [86]	2017	Argentina	Anfama	26°45′S, 65°34′W	22		CoproElisa	1	rural		1							7															
Parra et al. [86]	2017	Argentina	Chaquivil	26°41′S, 65°36′W	7		CoproElisa	1	rural		1							4															
Parra et al. [86]	2017	Argentina	La Hoyada	26°41′S, 65°31′W	5		CoproElisa	1	rural		1							3															
Parra et al. [86]	2017	Argentina	Mala Mala	26°47′S, 65°33′W	9		CoproElisa	1	rural		1							6															
Parra et al. [86]	2017	Argentina	San José de Chasquivil	26°41′S, 65°36′W	17		CoproElisa	1	rural		1							8															

Autor	Año	Country	Name Study Locality	Coordinates	Sample size	Fixing method	of tection Methods	No. Of detection methods	RURAL	URBAN	Richness	Ancylostomids	*Ancylostoma* sp.	*Uncinaria* sp.	*Ascaris* sp.	*Dibothriocephalus* sp.	*Dipylidium caninum*	*Echinococcus* sp.	*Eucoleus aerophila*	*Eucoleus boehmi*	Capillaria sp.	*Taenia multiceps* sp.	*Strongyloides* eggs	*Spirocerca*	*Taenidae*	*Taenia hydatigena*	*Taenia ovis*	*Toxascaris* sp.	*Toxocara* sp.	*Trichuris* sp.	Trematodes	*Oncicola canis*	*Physalopetra* sp.
Perez et al. [87]	2006	Argentina	Rio Negro	40°48′S, 63° 00′W	416		Copro, Elisa and WB	2			2							4.6							14.9								
Quilodrán-González et al. [88]	2018	Chile	Cabrero	37°2′S, 72°24′W	83		Flotation of Sheater	1		urban	1	41														4.8			4.8	13.3			
Quilodrán-González et al. [88]	2018	Chile	Cabrero	37°2′S, 72°24′W	10		Flotation of Sheater	1	rural		2	60																		10			
Radman et al. [89]	2006	Argentina	Capital Federal	34°34S, 58°31W	125		Flotation of Fülleborn	1		urban	1																		51.2				
Rivero et al. [90]	2015	Argentina	Puerto Iguazú y alrededres	25°35′S, 54°34′W	405	Formol 10%	Flotation of Sheater and Sedimentation of Telemann	2	rural		1					0.49																	
Rivero et al. [91]	2017	Argentina	Puerto Iguazú y alrededres	25°35′S, 54°34′W	530	Formol 10%	Direct with lugol, Flotation of Sheater and Sedimentation of Teleman	3		urban	8				0.9	0.9	1.3						55.6		0.4			3.9	13.4	12.1			
Rodriguez et al. [92]	2005	Argentina	Mar del Plata	38°00′S, 57°33′W	171		Flotation and Sedimentation	2		urban	6		67.8	42.4			1.5				5.6								6.8	52.2			
Roth et al. [93]	2018	Argentina	Bariloche	41°08′S, 71°27′W	118	Freezado	Flotation of Sheater and Sedimentation of Telemann	2		urban	1					16.9																	
Rubel et al. [94]	2003	Argentina	Capital Federal	34°34S, 58°31W	31	Formol 5%	Sedimentation of Teleman	1		urban	1																		14.0				
Rubel et al. [95]	2005	Argentina	Capital Federal	34°34S, 58°31′W	2417	Formol 5%	Sedimentation of Teleman	1		urban	4		33.5				0.7												13.0	32.0			
Rubel et al. [96]	2010	Argentina	Capital Federal	34°34′S, 58°31′W	421	Formol 5%	Flotation of Willis	1		urban	7		26.0				0.6						0.9		0.6			0.2	1.7	4.0			
Rubel et al. [97]	2019	Argentina	Buenos Aires	34°37′S, 58°25′W	112		Centrifugation and Flotation of Sheater	2		urban	4		20.5												0.9				1.8	3.6			
Sánchez et al. [98]	2003	Argentina	Comodoro Rivadavia y Rada Tilly	45°S, 68° W	481	Formol 5%	Sedimentation de Teleman and Flotation de Willis	2		urban	6			1.0			0.2				0.2			2.6	3.6				17.9				
Sánchez Thevenet et al. [99]	2003	Argentina	Comodoro Rivadavia	45°S, 68° W	163	Formol 5%	Sedimentation of Teleman and Flotation of Willis	2		urban	6			0.8			0.3							1.6	1.4				8.8				
Semenas et al. [100]	2014	Argentina	Bariloche	41°10′S, 71°18′W	54		Sedimentation of Teleman and Flotation of Sheater	2		urban	10		1.8	3.7		12.8		3.6			1.8		12.8		7.3			1.8	11.0	29.3			
Soriano et al. [101]	2010	Argentina	Neuquen rural	38°14′S, 69°46′W	1298	Formol 5%	Flotation and Sedimentation	2	rural		8		0.15			0.15	0.15								17.87			0.84	16.4	1.3		0.3	

Autor	Año	Country	Name Study Locality	Coordinates	Sample size	Fixing method	of tection Methods	No. Of detection methods	RURAL	URBAN	Richness	Ancylostomids	*Ancylostoma* sp.	*Uncinaria* sp.	*Ascaris sp.*	*Dibothriocephalus* sp.	*Dipylidium caninum*	*Echinococcus* sp.	*Eucoleus aerophila*	*Eucoleus boehmi*	Capillaria sp.	*Taenia multiceps* sp.	*Strongyloides eggs*	*Spirocerca*	*Taenidae*	*Taenia hydatigena*	*Taenia ovis*	*Toxascaris* sp.	*Toxocara* sp.	*Trichuris* sp.	Trematodes	*Oncicola canis*	*Physalopetra* sp.
Soriano et al. [101]	2010	Argentina	Neuquén urbano (neuquen y chos malal	37°23′S, 70°17′W	646	Formol 5%	Flotation and Sedimentation	2		urban	6		0.93								0.31				2.17				16.1	15.63			
Souto et al. [102]	2016	Argentina	El Chalía (Chubut)	45°41′S, 70°59′W	22	Formol 10%	Sedimentation of Teleman, Flotation of Willis and copro, Elisa	3	rural		2							13.6							9.1								
Taranto et al. [103]	2000	Argentina	Fortín Dragones y Misión Chaqueña	23°15′S, 63°20′W	106		Directo, Flotation of Willis and centrifugation	3		urban	4		69.8												1.9				17.2	7.5			
Torres et al. [104]	2004	Chile	Panguipulli	39°38′S, 72°20′W	109	PAF fenol, alcohol y formaldehído	Sedimentation	1		urban	1					1.8																	
Torres et al. [104]	2004	Chile	Choshuenco	39°50′S, 72°04′W	22	PAF fenol, alcohol y formaldehído	Sedimentation	1		urban	1					4.5																	
Vargas et al. [105]	2016	Chile	Niebla	39°48′S, 73°14′W	78	Formol salino	Sedimentation of Telemann modified, Flotation Sulphate Zinc, método cuantitativo	3			1																		15.4				
Vargas et al. [105]	2016	Chile	Valdivia	39°48′S, 73°14′W	77	Formol salino	Sedimentation of Telemann modified, Flotation Sulphate Zinc, método cuantitativo	3		urban	1																		15.6				
Winter et al. [106]	2018	Argentina	Viedma	40°48′S, 62°59′W	531		Flotation de Sheater	1		urban	6		33.8							2.2					0.7			2.9	22.8	40.4			
Zonta et al [107]	2019	Argentina	Clorinda (Formosa)	25°17′S, 57°43′W	16	Formol	Sedimentation of Ritchie and Flotation of Willis	2		urban	4		62.5	37.5																			
Zunino et al. [108]	2000	Argentina	Comodoro Rivadavia	45°S, 68° W	31	Formol 5%	Flotation of Willis	1		urban	2												9.7						3.3				
Zunino et al. [108]	2000	Argentina	Trelew	43°15′S, 65°18′W	30	Formol 5%	Flotation of Willis	1		urban	3			3.3									3.3						33.3				
Zunino et al. [108]	2000	Argentina	Puerto Madryn	42°46′S, 65°02′W	29	Formol 5%	Flotation of Willis	1		urban	1																		10.3				
Zunino et al. [108]	2000	Argentina	Sarmiento	45°36′S, 69°05′	29	Formol 5%	Flotation of Willis	1		urban	3			6.9									6.9						24.1				
Zunino et al. [108]	2000	Argentina	Esquel	42°54′S, 71°19′W	29	Formol 5%	Flotation of Willis	1		urban	3			6.9									3.4						13.8				
Zunino et al. [108]	2000	Argentina	Lago Puelo	42°09′S, 71°38′W	30	Formol 5%	Flotation of Willis	1		urban	3			16.7									6.9						20.0				

Table 1.
Data extracted from the 67 articles selected for analysis.

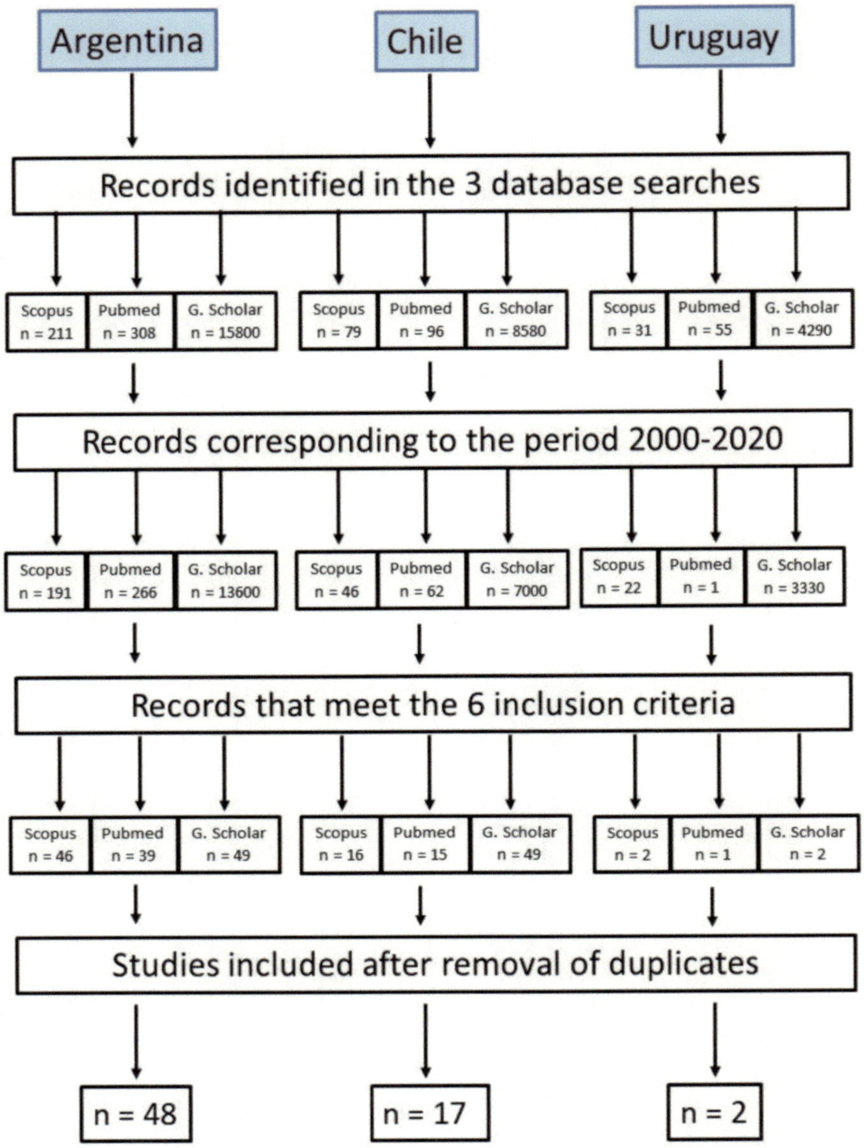

Figure 1.
Flow diagram of epidemiologic studies on dog parasites for the systematic review.

commonly used techniques were Willis flotation (20 reports), Sheater flotation (15 reports) and Telemann sedimentation (14 reports). In Uruguay only two methods were used: necropsy of stray dogs and coproELISA for *Echinococcus* sp., whereas in Argentina and Chile the techniques in common were Faust, Sheater, Telemann, Willis, and coproELISA for *Echinococcus* sp. Chilean researchers also used a modification of Faust (Teuscher), Burrows, and Harada-Mori for larvae. Other methods used only in Argentina were Füllerborn, Mini Flotac; Ritchie, Carles Barthelemy, direct observation with lugol; and Western Blot and PCR molecular techniques for *E. granulosus*.

More than 140 sites were analysed in Chile and Argentina (**Figure 2**, **Table 1**); however, the number of sites analysed in Uruguay could not be determined as this information is not given in the 2 selected studies. In Argentina and Chile, a total of 104 urban sites and 43 rural areas were considered (**Table 2**).

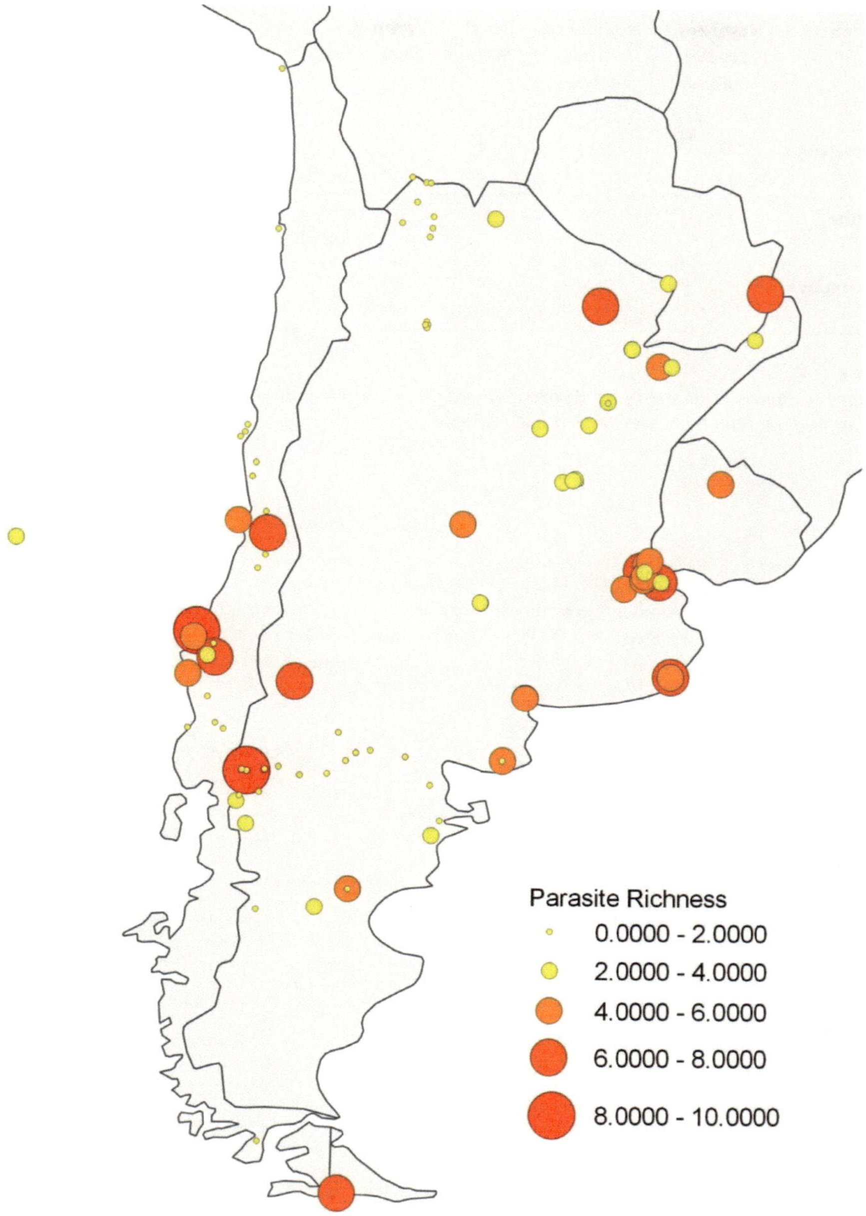

Figure 2.
Distribution of collection sites and species richness in each site.

A total of 22 parasite taxa was recorded (**Table 3**): 1 trematode (Trematoda sp.), 7 cestodes (*Dibothriocephalus* sp., *Dipylidium caninum*, *Echinococcus* sp., Taenidae, *Taenia multiceps*, *Taenia hydatigena*, *Taenia ovis*), 13 nematodes (*Trichuris vulpis, Eucoleus aerophila*, *Eucoleus boehmi*, *Capillaria* sp., *Strongyloides* sp., Ancylostomatidae sp., *Ancylostoma* sp., *Uncinaria* sp., *Ascaris* sp., *Toxascaris leonina*, *Toxocara canis, Spirocerca* sp., and *Physaloptera* sp.), and 1 acanthocephalan species (*Oncicola canis*). In Argentina the presence of *Ancylostoma* was recorded up to genus level, whereas in Chile they were recorded only as Ancylostomatidae sp., so while it is likely that there are some shared species, this cannot be established from the records analysed. The distribution of the species is presented in **Figures 3–5**, which show that most of the parasitic records are located in the central zone of Chile, while

Country	Number of studies analysed	Number of sites analysed	Rural Sites	Urban Sites	Total collected faeces (range)	Richness (Range)	Number of Techniques used
Argentina	48	110	38	76	18,812 (4–2417)	17 (1–10)	13
Chile	19	33	5	28	4,574 (10–972),	14 (1–9)	11
Uruguay	2	not reported	not reported	not reported	7,134 (79–5356)	6 (1–6)	2

Table 2.
Summary of studies: Total number of reports analysed for the three countries, number of rural and urban sites, collected samples, techniques used, and species richness.

Parasite species	Total Number of Sites	Mean prevalence (SD)	Number of positive urban sites	Mean prevalence in urban sites (SD)	Number of positive rural sites	Mean prevalence in rural sites (SD)
***Dibothriocephalus* sp.**	14	5.7 ± 6.2	10	7.8 ± 6.3	4	0.6 ± 0.4
Dipylidium caninum	21	5.6 ± 10.3	16	4.1 ± 9.3	5	10.5 ± 12.8
Echinococcus granulosus	52	7.9 ± 7.1	14	12.9 ± 9.9	38	6 ± 4.7
Taenidae	16	5.1 ± 5.9	12	3.4 ± 4.2	4	10.3 ± 7.5
Taenia hydatigena	9	9 ± 10.5	7	3.9 ± 2.7	2	26.8 ± 5.3
Taenia multiceps	2	2.5 ± 2.1	1	1	1	4
Taenia ovis	1	3			1	3
Trichuris vulpis	60	14.7 ± 14.7	53	15.3 ± 15.3	7	10.3 ± 8.7
Eucoleus aerophila	4	14.9 ± 8.8	1	17.4	3	14.1 ± 10.5
Eucoleus boehmi	2	1.8 ± 0.6	2	1.8 ± 0.6		
***Capillaria* sp.**	11	3.9 ± 6.1	11	3.9 ± 6.1		
***Strongyloides* sp.**	19	12 ± 16.1	14	5.6 ± 4.2	5	30.1 ± 22.7
Ancylostomatidae	6	24.2 ± 23.5	3	16 ± 21.7	3	32.3 ± 26.6
***Ancylostoma* sp.**	66	29 ± 23.4	62	29.7 ± 23.3	3	21.4 ± 27.2
***Uncinaria* sp.**	21	17.3 ± 18.5	17	18 ± 20.2	4	14.2 ± 8.8
***Ascaris* sp.**	8	7.6 ± 6.2	6	9.3 ± 6.1	2	2.5 ± 2.2
Toxascaris leonina	13	2.7 ± 3.2	11	2.7 ± 3.5	2	2.4 ± 2.2
Toxocara canis	86	13.6 ± 11.6	80	13.4 ± 11.5	6	15.9 ± 12.8
***Spirocerca* sp.**	3	3.4 ± 2.3	3	3.4 ± 2.3		
***Physalopetra* sp.**	1	1.2	1	1.2		
Oncicola canis	1	0.3			1	0.3

Table 3.
Species recorded in the studies analysed, their distribution (urban versus rural) and mean intensity.

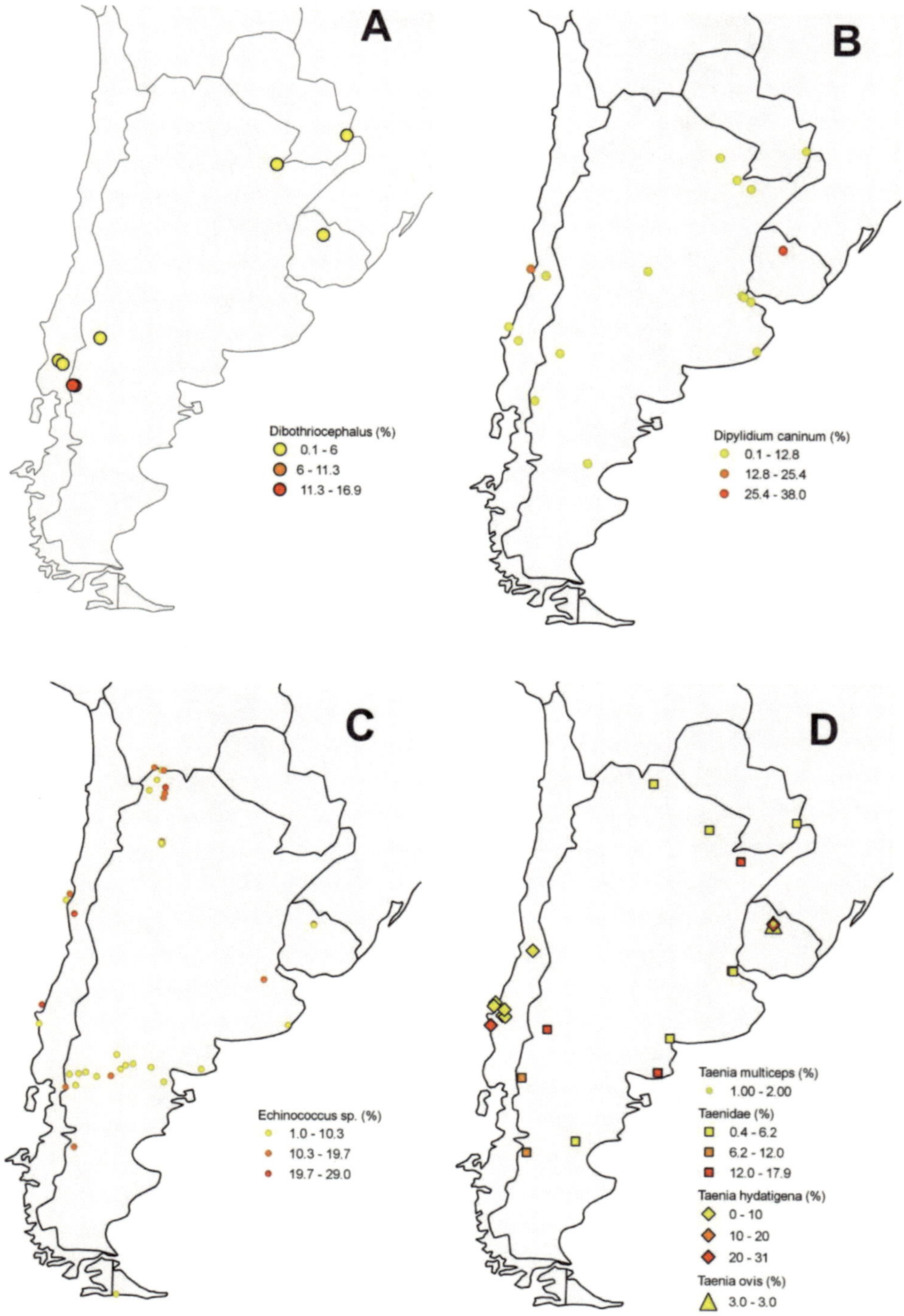

Figure 3.
Distribution of Cestoda collected in Argentina, Chile and Uruguay. A.: Dibothriocephalus *sp.; B.:* Dipilidium caninum*; C.:* Echinococcus *sp.; D.: Taenids.*

in Argentina there are records at all latitudes, except in an arid zone in the northwest, close to the Andes mountains. Species richness was correlated only with sample size ($R = 0.44809$, $p < 0.05$) and varied between sites, from 1 to 10 species (Argentina 1 to 10; Chile 1 to 9; Uruguay 1 to 6) (**Figure 2**).

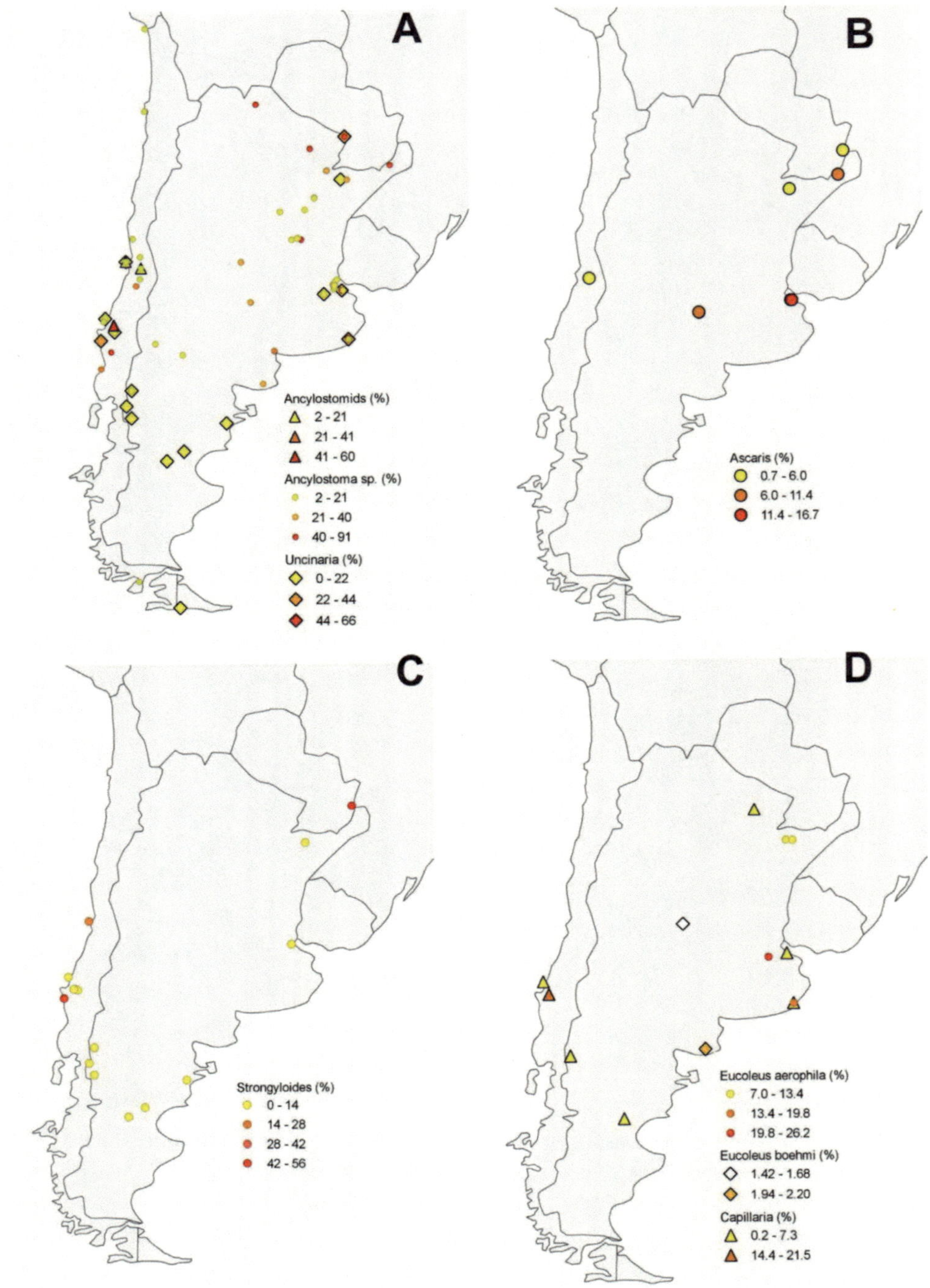

Figure 4.
Distribution of Nematoda (part 1) in Argentina, Chile and Uruguay. A.: Ancylostomatidae.; B.: Ascaris sp.; *C.:* Strongyloides; *D.:* Eucocleus *spp. and* Capillaria *sp.*

The most frequently recorded species was *T. canis* (86 sites), followed by *Ancylostoma* sp. (66); *Trichuris vulpis* (60 sites), and *Echinococcus* sp. (52) (**Table 3**; **Figure 4A, 5B, 3E**, respectively); others were recorded only once, e.g.: Trematoda sp. and *O. canis* in Argentina, and *Physaloptera* sp. in Chile. The species detected in Uruguay, except for *Echinococcus* sp., correspond to different taeniid cestodes. Argentina and Chile shared 10 helminth species: *Dibothriocephalus* sp., *D. caninum*

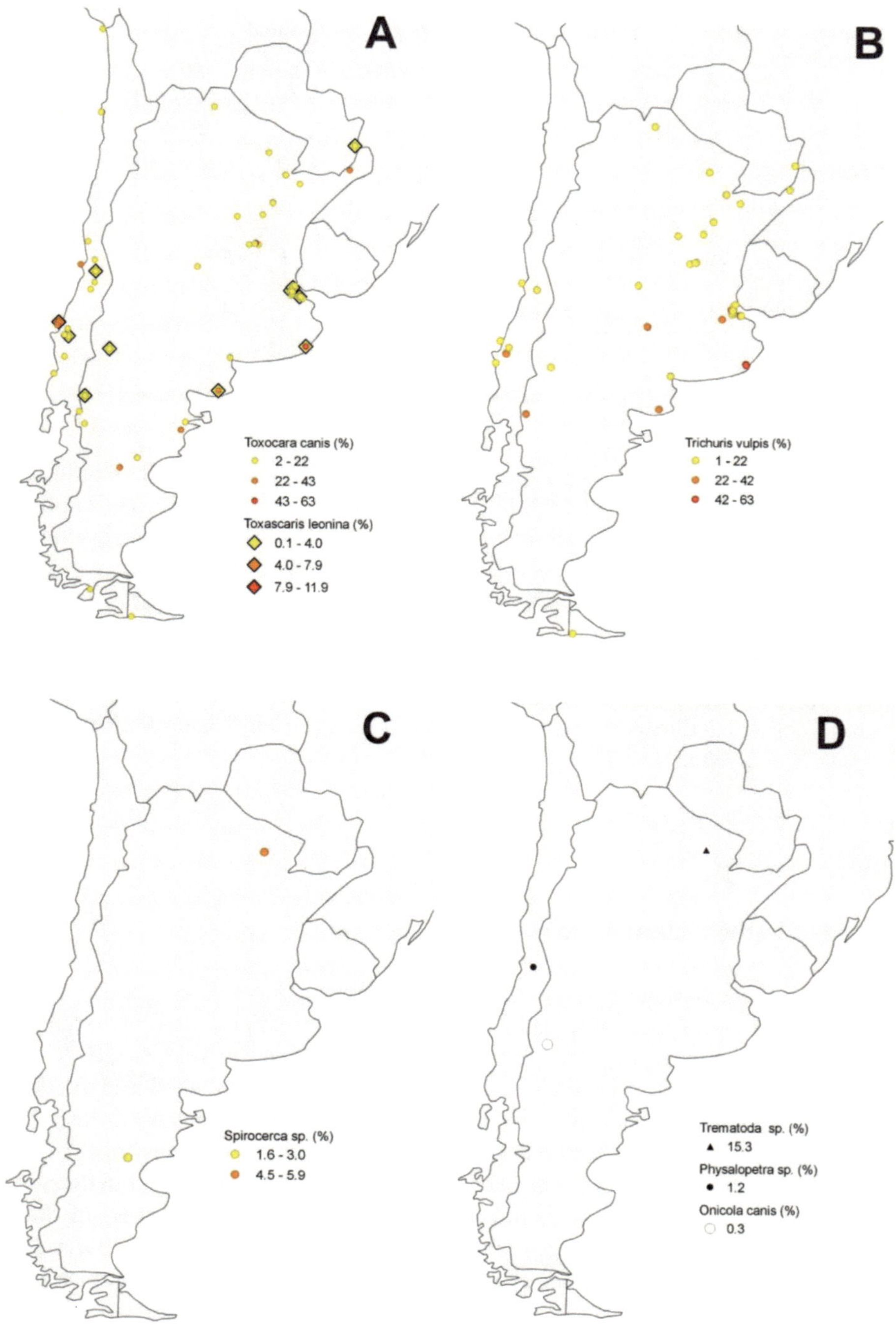

Figure 5.
Distribution of Nematoda (part 2) in Argentina, Chile and Uruguay. A.: Toxocara canis *and* Toxascaris leonina.*; B.:* Trichuris vulpis*; C.:* Spirocerca*; D.:* Physalopetra, *Trematoda* sp. *and* Oncicola canis.

sp., *Echinococcus* sp., *Ascaris* sp., *Capillaria* sp., *Strongyloides* sp., *T. leonina*, *T. canis*, *T. vulpis*, and *Uncinaria* sp.

The species richness in urban areas (20 species) was slightly higher than in rural areas (17 species) (**Table 3**). In addition, a higher number of zoonotic species was recorded in urban areas, species such as *Uncinaria* sp., *Ancylostoma* sp. and *Echinococcus* sp. being widespread and prevalent in the cities (**Table 3**). Many parasite

Country	Urban			Rural			
	Richness (Range)	*Mean richness*	*Most widespread species*	*Richness (Range)*	*Mean richness*	*Most widespread species*	*Similarity*
Argentina	16 (1–10)	3.8	*Toxocara canis*, *Ancylostoma* sp. *Trichuris vulpis*	10 (1–8)	1.7	*Echinococcus* sp.	7/17
Chile	14 (1–9)	2.8	*Toxocara canis*, *Ancylostoma* sp. *Trichuris vulpis*	8 (1–6)	3	*Echinococcus* sp.	7/14

Table 4.
Characterisation of urban and rural areas in terms of richness and most widespread species, present in Argentina and Chile.

species showed greater prevalence in urban areas than in rural ones. The only exception to this was *T. canis* which had higher values in the rural areas (**Table 3**). In Chile 8 species were registered in rural areas and 14 in urban locations, whereas in Argentina the species richness was 10 and 16, respectively (**Table 4**).

Of the total taxa recorded, 14 (63.6%) have been registered in humans: *Dibothriocephalus* sp., *D. caninum*, *Echinococcus* (*sensu lato*), Taenidae, *T. multiceps*, *T. hydatigena*, Ancylostomatidae sp., *Ancylostoma* sp., *Uncinaria* sp., *Ascaris* sp., *E. aerophila*, *E. boehmi*, *T. leonina,* and *T. canis.* Some of these species are only occasionally recorded infecting humans, such as *D. caninum*, *Taenia multiceps*, *E. aerophila*, *E. boehmi* and *T. leonina.*

4. Discussion

4.1 State of knowledge and distribution

Although three databases were used, this work could have some bias due to the exclusion of grey literature, like technical reports, congress abstracts or thesis manuscripts, so some sites or negative data may be excluded in the analysis [109]. The systematic bibliographic review carried out shows that the published and available knowledge of the occurrence and distribution of helminths in dogs is scarce in southern South America; in countries such as Uruguay there are no records other than those obtained within the Echinococcosis National Programmes. Furthermore, in Argentina there are arid regions near the Andes, such as the northwest of the country, where there are no records of parasites in dogs. The same was observed for Chile south to 40°s, except for one record in Punta Arenas, the southernmost city in Chile. Most of the records are associated with large cities and their surroundings, such as Buenos Aires and La Plata in Argentina, and in the area of Santiago de Chile, Concepción, and Temuco in Chile.

Although sample size is the only factor that significantly affected richness, other factors to consider could be the analytical methods used and whether the sample was fixed or not. Sample size affects the results, generating deviations in the number of species and in their prevalence, especially in places where the sample size was too low. On the other hand, a lack of methodological specifications can be observed in the techniques used. This could imply potential biases in the reporting and/or interpretation of data. In order to obtain data of higher quality, a general consensus should be reached on the techniques to be applied. It is also desirable to apply molecular techniques that allow parasite identification to species level, thus solving records identified to family level, such as "Ancylostomatidae" or "Strongylids", or the recording of species outside their natural range of distribution, like *Dibothriocephalus* in the northeast of Argentina.

The presence of a greater number of species, most of which have zoonotic potential, in urban areas than rural ones is probably due to the fact that dogs can roam freely. Dogs spread the parasite eggs, thereby these areas will function as contagion points for both other dogs and humans. A further problem is that deworming in these countries is insufficient [21]. A similar situation has been detected in parks in the United States, where it has been suggested that dogs are at risk of infection with parasites at these sites, and it has been recommended that preventive strategies be considered [30, 110]. Some parasitic infections could become increasingly urbanised, and an estimation for 2050 indicates that up to two-thirds of the global population will live in megacities. The slums of these megacities would concentrate high levels of intestinal helminth. Toxocariasis and other urban soil-transmitted helminths are important, yet little studied, health issues in the cities of the Americas [111].

The zoonotic broad tapeworm, *Dibothriocephalus* sp., is found in dogs from the endemic zone of the disease, the Andean Patagonia of Argentina and Chile [93, 104]. The records from the northeastern region of Argentina require revision, as there are no molecular studies confirming the identity of these parasites, and there are no records of fish infected by plerocercoids in this zone. Although *Dibothriocephalus* sp., is not transmitted to humans by dogs, they can act as disseminators of the disease and are often used as sentinel species for the spread of the disease in some areas. *Ascaris* sp. in dogs is distributed mainly in subtropical regions of Argentina, where this parasite is most prevalent in humans [107]. Some parasites are distributed throughout all the latitudes regardless of the type of climate, like *T. canis*., *T. vulpis*, and Ancylostomatids, as observed in other parts of the world [112–114]. *Echinococcus* sp. is distributed across almost all rural areas of the three countries, although has recently also been registered in cities [35, 47, 64, 115].

4.2 Zoonoses and human cases reported

The high percentage of parasites with zoonotic potential reinforces the need to establish effective prevention measures, not only with regard to parasitosis in animals but also to transmission to humans. This situation highlights the need for better integration between specialists in animal and human health [74]. A few diseases transmitted by dogs have surveillance mechanisms in humans, but there are many other important zoonoses worldwide, with numerous human cases, which are not kept watch on. Some of these have been recorded in Argentina and Chile, such as those caused by *T. canis*, *Ancylostoma* sp., *A. caninum*, *Uncinaria* sp., and *Strongyloides* sp. [30]. Of the main zoonoses recorded in dogs in the three countries, cystic echinococcosis is the only one which has to be reported to the health authorities, since it is of major sanitary importance [115]. The others, like toxocariasis, hookworm and strongyloidiasis are not reported, and records of human cases in these countries are scarce. The status of these zoonoses in humans from southern South America is analysed below.

4.2.1 Cystic echinococcosis

Cystic echinococcosis or hydatidosis, produced by *Echinococcus granulosus sensu lato*, is a highly endemic parasitic zoonosis in South American countries, especially in Argentina, Chile, Uruguay and Brazil. It is associated with rural areas dedicated mainly to goat and sheep breeding, and causes significant economic losses [47, 69, 116–118].

From 2009 to 2014, a total of 29,559 new human cases of cystic echinococcosis were registered in these countries. The average fatality rate across the three countries was 2.9%, suggesting that the disease causes approximately 880 deaths annually. The most affected are children <15 years of age, which is indicative of a persistent environmental risk leading to new cases [69, 115]. In the countries analysed, Government Control Programmes have been addressed, and surveillance of the disease from a holistic perspective based on Primary Health Care has been implemented [64, 69, 115, 117]. The number of human cases has a heterogeneous geographical distribution in Chile and Argentina, showing an increase towards the south [116, 118].

4.2.2 Toxocariasis

Toxocariasis is an infection that has a worldwide distribution and is a very important zoonosis due to its frequent occurrence in humans [119]. The estimate of the overall worldwide prevalence of *T. canis* in dogs of 11.1% represents 100 million dogs, which should alert Public Health experts and policy makers to the need for effective intervention programs [114, 120]. This parasite species has high biotic potential since its eggs contaminate water, soil, grass, and pet fur [51]. The results presented here regarding *T. canis* in dogs of southern South America show higher prevalence values (around 13%) than the overall prevalence registered worldwide. Also, the risk of infection is similar in urban and rural areas, as suggested in Chile [105]. In Argentina, numerous studies that analysed the seroprevalence of toxocariasis in both children and adults from urban and rural areas reported results varying between 28% and 80% [51, 121, 122]. In Chile, the seroprevalence of this parasitosis varies between 1.3% and 25.4% [105]. Although in Uruguay there are no published records of seroprevalence in humans [123], a recently published work reported that from 2014 to 2018, 20 children had been treated in the public health system for ocular and visceral *larva migrans* syndrome [123].

4.2.3 Ancylostomiasis

Dog hookworms are *Ancylostoma caninum*, *Ancylostoma braziliense*, and *Uncinaria stenocephala*, and their eggs can be found in faeces. The larvae of these parasites can cause cutaneous *larva migrans* in humans [124]. The main causal agent of *larva migrans* worldwide is *A. braziliense*; however, the causative agents vary among geographical areas, even within a single country. This disease is mainly endemic to tropical and subtropical developing countries with high average annual temperatures and humid climates, predominating in America from the southern United States, through Mexico, Central, and reaching South America. It is especially prevalent in areas where dogs roam freely, and on sandy, wet soils, such as beaches and playgrounds [124]. In Argentina, records of human cutaneous *larva migrans* correspond to the *Wichi* aboriginal communities in the subtropics of the northwest of the country [103], or to people who had travelled to Brazil [125]. In Chile, there are also few reports of this disease, and they correspond to a 3-year-old patient who acquired the disease in an urban area [126], and to an adult who had been infected on a trip to Brazil [127].

4.2.4 Strongyloidiasis

Strongyloidiasis is prevalent in remote socioeconomically disadvantaged communities around the world, and dogs can act as reservoirs of human strongyloidiasis [128]. This parasitosis is registered in the north of Argentina, with similar infection values in both rural and urban populations and an overall seroprevalence of 19.6%

[129, 130]. In Chile, the seroprevalence is much lower (0.25%) in blood donors from Arica and La Union. Human infections by *S. stercoralis* in this country are therefore endemic, with very low frequency in apparently healthy individuals [131].

5. General conclusions

This review shows that knowledge of canine helminths in southern South America is scarce. The studies published on dog parasites are not equally distributed across the three countries, with Uruguay presenting the least amount of available information. Data on dog parasites in southern South America is still too incipient for identification of a clear distribution pattern. Homogenisation of criteria would be beneficial, since the methods used are diverse and heterogeneous, some studies using only flotation or sedimentation techniques. Numerous parasitic species were recorded, many of which are zoonotic and widely distributed throughout both urban and rural areas of these countries. The risk of dogs becoming infected is high given the number of parasites present and the style of pet ownership in the communities of these countries, where dogs are allowed to roam freely, and veterinary care is scarce. The high percentage of zoonotic helminths reinforces the need to establish effective prevention measures, not only for parasitosis in animals but also for transmission to humans. Considering that people in both urban and rural areas are at risk of being infected with zoonoses transmitted by dogs, given the high levels of infection they present in their faeces, a One Health approach to public health would be desirable, such that humans and dogs should be treated concomitantly to control the parasites. Furthermore, it would be desirable to implement measures such as control of the canine population, mass treatment of dogs with anthelmintics, education programmes and healthcare alert systems.

Acknowledgements

This work was funded by PICT 1385–2017 and UNCo B225.

Author details

Luciano Ritossa, Gustavo Viozzi and Verónica Flores*
Laboratorio de Parasitología, Instituto de Investigaciones en Biodiversidad y Medioambiente (Consejo Nacional de Investigaciones Científicas y Técnicas - Universidad Nacional del Comahue), San Carlos de Bariloche, Río Negro, Argentina

*Address all correspondence to: veronicaroxanaflores@gmail.com

References

[1] Germonpré M, Sablin M, Stevens R, Hedges R, Hofreiter M, Stiller M, Despres V. Fossil dogs and wolves from Paleolithic sites in Belgium, the Ukraine and Russia: osteometry, ancient DNA and stable isotopes. Journal of Archaeological Science. 2009; 36:473–490. DOI: 10.1016/j.jas.2008.09.033

[2] Morand S, McIntyre KM, Baylis M. Domesticated animals and human infectious diseases of zoonotic origins: Domestication time matters. Infection, Genetics and Evolution. 2014; 24:76–81. DOI: 10.1016/j.meegid.2014.02.013

[3] O'Haire M. Companion animals and human health: Benefits, challenges, and the road ahead. Journal of Veterinary Behavior. 2010; 5:226–234. DOI: 10.1016/j.jveb.2010.02.002

[4] Peña A, Abarca K, Weitzel T, Gallegos J, Cerda J, García P, López J. One Health in Practice: a pilot project for integrated care of zoonotic infections in immunocompromised children and their pets in Chile. Zoonoses and Public Health. 2015; 63: 403–409. DOI:10.1111/zph.12241

[5] Hill J, Ziviani J, Driscoll C, Teoh A, Chua J, Cawdell-Smith J. 2020. Canine assisted occupational therapy for children on the autism spectrum: A pilot randomized control trial. Review Journal of Autism and Developmental Disorders. 2020; 50:4106–4120. DOI: 10.1007/s10803-020-04483-7

[6] Cleaveland SC, Laurenson MK, Taylor LH. Diseases of humans and their domestic mammals; pathogen characteristics, host range and the risk of emergence. Philosophical Transactions of the Royal Society of London. 2001; 356:991–999. DOI: 10.1098/rstb.2001.0889

[7] Gompper ME. Free-ranging dogs and wildlife conservation. Oxford and New York: Oxford University Press; 2013. 312 p.

[8] Wellbeing international NGO [Internet]. Available from: https://wellbeingintl.org/global-dog-populations-2/ [Accessed: 2020-12-08]

[9] World Health Organization [Internet]. Available from: https://www.who.int/es [Accessed: 2020-12-08]

[10] Carvelli A, Scaramozzino P, Iacoponi F, Condoleo R, Ugo Della M. Size, demography, ownership profiles, and identification rate of the owned dog population in central Italy. PLoS ONE. 2020; 15: e02405515. DOI: 10.1371/journal.pone.0240551

[11] Insurance Information Institution [Internet]. Available from: https://www.iii.org/table-archive/22742 [Accessed: 2020-12-10]

[12] Otranto D, Dantas-Torres F, Mihalca AD, Traub RJ, Lappin M, Baneth G. Zoonotic parasites of sheltered and stray dogs in the era of the global economic and political crisis. Trends in Parasitology. 2017; 33:813–825. DOI: 10.1016/j.pt.2017.05.013

[13] Infobae [Internet]. 2011. Available from: https://www.infobae.com/2011/09/16/605985-argentina-el-pais-la-region-mas-mascotas-habitante/#:~:text=Argentina%20es%20el%20pa%C3%ADs%20de,%25 [Accessed: 2020-12-14]

[14] Astorga F, Escobar LE, Poo-Muñoz DA, Medina-Vogel G. Dog ownership, abundance and potential for bat-borne rabies spillover in Chile. Preventive Veterinary Medicine. 2015; 118:397–405. DOI: 10.1016/j.prevetmed.2015.01.002

[15] El Observador [Internet]. 2017. Available from: https://www.elobservador.com.uy/nota/-cuantos-perros-y-gatos-hay-en-uruguay–201751716300 [Accessed: 2020-12-15]

[16] Zanini F, Leiva D, Fernández R, Bergagna H, Elissondo MC. Manejo de las poblaciones caninas urbanas en Argentina. Revista Argentina de Zoonosis y Enfermedades Infecciosas Emergentes. 2013; 8:20–25. aazoonosis.org.ar/wp-content/uploads/2013/05/RAZ-2013-Vol-VIII.pdf

[17] De la Reta M, Muratore M, Perna S, Polop J, Provensal MC. Abundancia de perros en situación de calle y su relación con factores ambientales en Río Cuarto (Córdoba, Argentina). Revista Veterinaria. 2018; 29:113–118. DOI: 10.30972/vet.2923275

[18] Notejane M, Moure T, Silva JED, Barrios P, Pérez W. Niños con mordeduras de animales hospitalizados en un centro de referencia de Uruguay. Boletín Médico del Hospital Infantil de México. 2018; 75:358–365. DOI: 10.24875/bmhim.18000031

[19] Schüttler E, Saavedra-Aracena L, Jiménez JE. Domestic carnivore interactions with wildlife in the Cape Horn Biosphere Reserve, Chile: husbandry and perceptions of impact from a community perspective. PeerJ. 2018; 6:e4124. DOI: 10.7717/peerj.4124

[20] Natalini B, Gennuso S, Beldoménico PM, Rigonatto T, Kowalewski MM. Parasitologic examination and associated risk factors of domestic dogs at the domestic-wildlife interface in the Iberá wetlands Ecoregion, Argentina. Veterinary Parasitology: Regional Studies and Reports. 2020; 20:100378. DOI: 10.1016/j.vprsr.2020.100378

[21] Astorga F, Poo-Muñoz DA, Organ J, Medina-Vogel G. Why let the dogs out? Exploring variables associated with dog confinement and general characteristics of the free-ranging owned-dog population in a peri-urban area. Journal of Applied Animal Welfare Science. 2020; 27:1–15. DOI: 10.1080/10888705.2020.1820334

[22] López J, Abarca K, Cerda J, Valenzuela B, Lorca L, Olea A, Aguilera X. Surveillance system for infectious diseases of pets, Santiago, Chile. Emerg ing Infectious Disease. 2009; 15:1674–1676. DOI: 10.3201/eid1510.081596

[23] Garibotti G, Zacharías D, Flores V, Catrimán S, Falconaro A, Kabaradjian S, Luque ML, Macedo B, Molina J, Rauque C, Soto M, Vázquez G, Vega R, Viozzi G. Tenencia responsable de perros y salud humana en barrios de San Carlos de Bariloche, Argentina. Medicina. 2017; 77:309–313. http://www.medicinabuenosaires.com/revistas/vol77-17/n4/309-313-Med6681-Garibotti.pdf

[24] Canadian guidelines for the treatment of parasites in dogs and cats [Internet]. 2009. Available from: https://www.wormsandgermsblog.com/files/2008/03/CPEP-guidelines-ENGLISH1.pdf [Accessed: 2020-12-14]

[25] Villagra V, Cáceres D, Alvarado S, Salinas E, Caldera ML, Lucero E, Viviani P, Torres M. Características epidemiológicas de mordeduras en personas, según registro de atención de urgencias Provincia de Los Andes, Chile. Revista Chilena de Infectología. 2017; 34:212–220. DOI: 10.4067/S0716-10182017000300002

[26] Pro Tenencia - Tenencia Responsable de perros y gatos en Argentina. [Internet]. Available from: https://www.argentina.gob.ar/salud/protenencia. [Accessed: 2020-12-10]

[27] Intendencia de Montevideo [Internet]. 2020. Available from: https://montevideo.gub.uy/noticias/salud-y-salubridad/calendario-de-castraciones-de-perros-y-gatos [Accessed: 2020-12-14]

[28] Gobierno de la ciudad de Buenos Aires. Informe módulo de Tenencia responsable y sanidad de perros y gatos. Encuesta Anual de Hogares 2018 Ciudad de Buenos Aires. [Internet]. 2020.

Available from: https://www.estadisticaciudad.gob.ar/eyc/wp content/uploads/2020/01/eah_2018_tenencia_responsable_perros_gatos.pdf [Accessed: 2020-12-10]

[29] Andresiuk MV, Rodríguez F, Denegri GM, Sardella NH, Hollmann P. Relevamiento de parásitos zoonóticos en materia fecal canina y su importancia para la salud de los niños. Archivos Argentinos de Pediatría. 2004; 102:325–329. https://www.sap.org.ar/docs/publicaciones/archivosarg/2004/A5.325-329.Andresiuk.pdf

[30] Day MJ, Breitschwerdt E, Cleaveland S, Karkare U, Khanna C, Kirpensteijn J. Surveillance of zoonotic infectious disease transmitted by small companion animals. Emerging Infectious Diseases. 2012; 18:e1. DOI: 10.3201/eid1812.120664

[31] Gorman T, Soto A, Alcaíno H. Parasitismo gastrointestinal en perros de comunas de Santiago de diferente nivel socioeconómico. Parasitología Latinoamericana. 2006; 61:126–132. DOI: 10.4067/S0717-77122006000200005

[32] Cociancic P, Deferrari G, Zonta ML, Navone GT. Intestinal parasites in canine feces contaminating urban and recreational areas in Ushuaia (Argentina). Veterinary Parasitology: Regional Studies and Reports. 2020; 21: 100424. DOI: 10.1016/j.vprsr.2020.100424

[33] López-Pérez AM, Orozco L, Zazueta OE, Fierro M, Gomez P, Foley J. An exploratory analysis of demography and movement patterns of dogs: New insights in the ecology of endemic Rocky Mountain-Spotted Fever in Mexicali, México. PLoS One. 2020; 15: e0233567. DOI: 10.1371/journal.pone.0233567

[34] Sepúlveda MA, Singer RS, Silva-Rodríguez E, Stowhas P, Pelican K. Domestic dogs in rural communities around protected areas: Conservation problem or conflict solution? PLoS ONE. 2014; 9:e86152. DOI: 10.1371/journal.pone.0086152

[35] Flores V, Viozzi G, Garibotti G, Zacharias D, Debiaggi MF, Kabaradjian S. Echinococcosis and other parasitic infection in domestic dogs from urban areas of an Argentinian Patagonian city. Medicina. 2017; 77:469–474. http://www.medicinabuenosaires.com/PMID/29223937.pdf

[36] Arezo M, Mujica G, Uchiumi L, Santillán G, Herrero E, Labanchi JL, Arayaa D, Salvitti JC, Cabrera M, Grizmado C, Calabro A, Talmon G, Sepulveda L, Galvan JM, Volpe M, Bastin V, Seleiman M, Panomarenko O, Tissot H, Sobrino M, Crowley P, Daffner J, Larrieu E. Identification of potential 'hot spots' of cystic echinococcosis transmission in the province of Río Negro, Argentina. Acta Tropica. 2020; 204:105341. DOI: 10.1016/j.actatropica.2020.105341

[37] Little SE, Johnson EM, Lewis D, Jaklitsch RP, Payton ME, Blagburn BL, Bowmane DD, Moroff S, Tams T, Rich L, Aucoin D. Prevalence of intestinal parasites in pet dogs in the United States. Veterinary Parasitology. 2009; 166:144–152. DOI: 10.1016/j.vetpar.2009.07.044

[38] Deplazes P, Van Knapen F, Schweiger A, Overgaauw PAM. Role of pet dogs and cats in the transmission of helminthic zoonoses in Europe, with a focus on echinococcosis and toxocarosis. Veterinary Parasitology. 2011; 182:41–53. DOI: 10.1016/j.vetpar.2011.07.014

[39] Areekul P, Putaporntip C, Pattanawong U, Sitthicharoenchai P, Jongwutiwes S. *Trichuris vulpis* and *T. trichiura* infections among schoolchildren of a rural community in north-western Thailand: the possible role of dogs in disease transmission.

Asian Biomedicine. 2010; 4:49–60. DOI: 10.2478/abm-2010-0006

[40] Shalaby HA, Abdel-Shafy S, Derbala AA. The role of dogs in transmission of *Ascaris lumbricoides* for humans. Parasitology Research. 2010; 106:1021–1026. DOI: 10.1007/s00436-010-1755-8

[41] Štrkolcová G, Goldová M, Bocková E, Mojžišová J. The roundworm *Strongyloides stercoralis* in children, dogs, and soil inside and outside a segregated settlement in Eastern Slovakia: frequent but hardly detectable parasite. Parasitology Research. 2017; 116:891–900. DOI: 10.1007/s00436-016-5362-1

[42] Amundson Romich J. Understanding Zoonotic Diseases. Nueva York: Thomson Delmar Learning; 2008. 701 p.

[43] Dantas Torres F, Otranto D. Dogs, cats, parasites, and humans in Brazil: opening the black box. Parasites & Vec tors. 2014; 7:22. DOI: 10.1186/1756-3305-7-22

[44] Alvarez Di Fino EM, Rubio J, Abril MC, Porcasi X, Periago MV. Risk map development for soil-transmitted helminth infections in Argentina. PLOS Neglected Tropical Diseases. 2020; 14: e0008000. DOI: 10.1371/journal.pntd.0008000

[45] Aromantaris E, Fernández R, Godfrey CM, Holly C, Khalil H, Tungpunkom P. Summarizing systematic reviews: methodological development, conduct and reporting of an umbrella review approach. Interna tional Journal of Evidence-Based Hea lthcare. 2015; 13:132–140. DOI: 10.1097/XEB.0000000000000055

[46] Instituto Geográfico Nacional [Internet]. Available from: https://www.ign.gob.ar. [Accessed: 2020-12-10].

[47] Acosta-Jamett G, Cleaveland S, Bronsvoort BM, Cunningham AA, Bradshaw H, Craig PS. *Echinococcus granulosus* infection in domestic dogs in urban and rural areas of the Coquimbo region, north-central Chile. Veterinary Parasitology. 2010; 19:117–22. DOI: 10.1016/j.vetpar.2009.12.005.

[48] Acosta-Jamett G, Weitzel T, Boufana B, Adones C, Bahamonde A, Abarca K, Craig PS, Reiter-Owona I. Prevalence and risk factors for echinococcal infection in a rural area of northern Chile: a household-based cross-sectional study. PLoS Neglected Tropical Disease. 2014; 8:e3090. DOI: 10.1371/journal.pntd.0003090.

[49] Andresiuk MV, Sardella NH, Denegri GM. Seasonal fluctuations in prevalence of dog intestinal parasites in public squares of Mar del Plata city, Argentina and its risk for humans. Revista Argentina de Microbiología. 2007; 39:221–224. https://www.redalyc.org/pdf/2130/213016793007.pdf

[50] Andresiuk MV, Denegri GM, Sardella NH, Hollmann P. Encuesta coproparasitológica canina realizada en plazas públicas de la ciudad de Mar del Plata, Buenos Aires, Argentina. Parasitología Latinoamericana. 2003; 58: 17–22. DOI: 10.4067/S0717-77122003000100003.

[51] Archelli S, Kozubsky L, Gamboa MI, Osen B, Costas ME, López M, Burgos L, Corbalán V, Butti M, Radman N. *Toxocara canis* en humanos, perros y suelos en ribera del Río de la Plata, provincia de Buenos Aires. Acta de Bioquímica Clínica Latinoamericana. 2018; 52:441–449. http://sedici.unlp.edu.ar/bitstream/handle/10915/73806/Documento_completo.pdf-PDFA.pdf?sequence=1&isAllowed=y

[52] Armstrong WA, Oberg C, Orellana JJ. Presence of parasite eggs with zoonotic potential in parks and public squares of the city of Temuco, Araucanía Region, Chile. Archivos de Medicina Veterinaria. 2011; 43:127–134.

DOI: 10.4067/S0301-732X2011000200005.

[53] Casas N, Costas S, Céspedes G, Sosa S, Santillán G. Detección de coproantígenos para el diagnóstico de equinococosis canina en la zona fronteriza de La Quiaca-Villazón. Revista Argentina de Microbiología. 2013; 45:154–159. DOI: 10.1016/S0325-7541(13)70017-8.

[54] Castillo D, Paredes C, Zañartu C, Castillo G, Mercado R, Muñoz V, Schenone H. Environmental contamination with *Toxocara* sp. eggs in public squares and parks from Santiago, Chile, 1999. Boletín Chileno de Parasitología. 2000; 55:86–91. DOI: 10.4067/S0716-10182016000400007

[55] Chiodo P, Basualdo J, Ciarmela L, Pezzani B, Apezteguía M, Minvielle M. Related factors to human toxocariasis in a rural community of Argentina. Memórias do Instituto Oswaldo Cruz, Rio de Janeiro. 2006; 101:397–400. DOI: 10.1590/S0074-02762006000400009.

[56] Cociancic P, Zonta ML, Navone GT. A cross-sectional study of intestinal parasitoses in dogs and children of the periurban area of La Plata (Buenos Aires, Argentina): Zoonotic importance and implications in public health. Zoonoses Public Health. 2017; 65:44–53. DOI: 10.1111/zph.12408.

[57] de Costas S, Riveros Matas N, Ricoy G, Sosa S, Santillán G. Diagnóstico de situación de la equinococosis quística en heces dispersas en las zonas de Quebrada y Puna, provincia de Jujuy, Argentina. Revista Argentina de Microbiología. 2014; 46:80–84. DOI: 10.1016/S0325-7541(14)70052-5

[58] Dopchiz MC, Lavallén CM, Bongiovanni R, Gonzalez PV, Elissondo C, Yannarella F, Denegri G. Endoparasitic infections in dogs from rural areas in the Lobos District, Buenos Aires province, Argentina. Revista Brasileira de Parasitologia Veterinária. 2013; 22:92–97. DOI: 10.1590/S1984-29612013005000008.

[59] Enriquez GF, Macchiaverna NP, Argibay HD, López Arias L, Farber M, Gürtler RE, Cardinal MV, Garbossa G. Polyparasitism and zoonotic parasites in dogs from a rural area of the Argentine Chaco. Veterinary Parasitology: Regional Studies and Reports. 2019; 16: 100287. DOI: 10.1016/j.vprsr.2019.100287.

[60] Fontanarrosa M, Vezzani D, Basabe J, Eira D. An epidemiological study of gastrointestinal parasites of dogs from Southern Greater Buenos Aires (Argentina): Age, gender, breed, mixed infections, and seasonal and spatial patterns. Veterinary Parasitology. 2006; 136:283–295. DOI: 10.1016/j.vetpar.2005.11.012.

[61] Gamboa MI, Navone GT, Orden AB, Torres MF, Castro LE, Oyhenart EE. Socio-environmental conditions, intestinal parasitic infections and nutritional status in children from a suburban neighborhood of La Plata, Argentina. Acta Tropica. 2011; 118:184–189. DOI: 10.1016/j.actatropica.2009.06.015.

[62] Gamboa MI, Kozubsky LE, Costas ME, Garraza M, Cardozo MI, Susevich ML, Magistrello PN, Navone GT. Asociación entre geohelmintos y condiciones socioambientales en diferentes poblaciones humanas de Argentina. Revista Panamericana de la Salud Pública. 2009; 26:1–8. https://scielosp.org/article/rpsp/2009.v26n1/1-8/

[63] González-Acuña D, Moreno L, Hermosilla C. Parásitos en perros de San Juan Bautista, Isla Robinson Crusoe, Chile. Archivos de Medicina Veterinaria. 2008; 40:193–195. DOI: 10.4067/S0301-732X2008000200012.

[64] Irabedra P, Ferreira C, Sayes J, Elola S, Rodríguez M, Morel N, Segura S, Dos

Santos E, Guisantes JA. Control programme for cystic echinococcosis in Uruguay. Memórias do Instituto Oswaldo Cruz, Rio de Janeiro. 2016; 111: 372–377. DOI: 10.1590/0074-02760160070.

[65] La Sala LF, Leiboff A, Burgos JM, Costamagna SR. Spatial distribution of canine zoonotic enteroparasites in Bahía Blanca, Argentina. Revista Argentina de Microbiología. 2015a; 47:17–24. DOI: 10.1016/j.ram.2014.12.006.

[66] La Sala LF, Costamagna SR, Leiboff A. Zoonotic parasites in dog feces in Bahía Blanca city. Revista Argentina de Zoonosis y Enfermedades Infecciosas Emergentes. 2015b; 10:70–71. https://www.aazoonosis.org.ar/wp-content/uploads/2013/05/Zoo-2015-2-completa-ok.pdf

[67] Lamberti R, Gino L, Larrieu E, García Cachau M, Calvo C, Morete M, Molina L, Lapuyade C, Cornejo T, Poblete G, Baeza R, Arias P, Cuellas F, Berrios Sierpe A, Crivelli L, Cejas C. Contaminación de parásitos zoonóticos en espacios públicos en el área del Centro de Salud Brown, General Pico, La Pampa. Comunicación preliminar. Revista Ciencias Veterinarias. 2014; 16: 57–65. https://repo.unlpam.edu.ar/handle/unlpam/4410

[68] Lamberti R, Gino L, García Cachau M, Calvo C, Morete M, Lapuyade C, Molina L, Larrieu E, Santos K, Arias P, Cornejo T. Contaminación de espacios públicos con nematodes zoonóticos en el Área del Centro de Salud Brown, General Pico, La Pampa desde 2013 a 2015. Ciencia Veterinaria. 2015; 17:61–71. DOI: 10.19137/cienvet20141614

[69] Larrieu E, Seleiman M, Herrero E, Mujica G, Labanchi JL, Araya D, Grizmado C, Sepúlveda L, Calabro A, Talmón G, Crowley P, Albarracín S, Arezo M, Volpe M, Ávila A, Pérez A, Uchiumi L, Salvitti JC, Santillán G. Vigilancia de la equinococosis quística en perros y niños en la provincia de Río Negro, Argentina. Revista Argentina de Microbiología. 2014; 46:91–97. DOI: 10.1016/S0325-7541(14)70054-9.

[70] Lavallén CM, Dopchiz MC, Lobianco E, Hollmann P, Denegri G. Intestinal parasites of zoonotic importance in dogs from the District of General Pueyrredón (Buenos Aires, Argentina). Revista Veterinaria. 2011; 22:19–24. https://revistas.unne.edu.ar/index.php/vet/article/view/19/14

[71] López J, Abarca K, Paredes P, Inzunza E. Intestinal parasites in dogs and cats with gastrointestinal symptoms in Santiago, Chile. Revista Médica de Chile. 2006; 134:193–200. DOI: 10.4067/s0034-98872006000200009.

[72] Luzio A, Espejo S, Troncoso I, Fernández I, Fischer C. Determinación coproscópica de formas parasitarias en heces de "*Canis lupus familiaris*" diseminadas en playas de la comuna de Tomé, Región del Bío Bío, Chile. Revista Ibero-Latinoamericana de Parasitología. 2013; 72:88–94. https://www.redalyc.org/pdf/636/63653009042.pdf

[73] Luzio Á, Belmar P, Troncoso I, Luzio P, Jara A, Fernández Í. Parasites of zoonotic importance in dog feces collected in parks and public squares of the city of Los Angeles, Bío-Bío, Chile. Revista Chilena de Infectología. 2015; 32:403–407. DOI: 10.4067/s0716-10182015000500006.

[74] Luzio A, Díaz P, Luzio P, Fernández I. Parasites of zoonotic importance in dog feces collected in square of arms of the provincial capitals of the Bío Bío Region, Chile. Revista Electrónica de Veterinaria. 2017; 18:1–10. https://www.redalyc.org/pdf/636/63653009042.pdf

[75] Madrid V, Sardella N, Hollmann P, Denegri G. Estudio coproparasitológico canino en playas de Mar del Plata y su impacto en la salud pública. Revista Veterinaria. 2008; 19:23–27. DOI: 10.30972/vet.1914294.

[76] Marder G, Ulon S, Botinelli O, Mesa Fleitas Z, Lotero DA, Ruiz R, Peiretti HA, Arzú RA. Infestación parasitaria en suelos y materia fecal de perros y gatos de la ciudad de Corrientes. Revista Veterinaria. 2004; 15:70–72. https://revistas.unne.edu.ar/index.php/vet/article/view/1999

[77] Martín UO, Demonte MA. Urban contamination with zoonotic parasites in the central region of Argentina. Medicina. 2008; 68:363–366. http://medicinabuenosaires.com/revistas/vol68-08/5/v68_n5_p363_366_.pdf

[78] Mercado R, Ueta MT, Castillo D, Muñoz V, Schenone H. Exposure to *larva migrans* syndromes in squares and public parks of cities in Chile. Revista de Saúde Pública. 2004; 38:729–731. DOI: 10.1590/S0034-89102004000500017

[79] Milano AMF, Oscherov EB. Contaminación de aceras con enteroparásitos caninos en Corrientes, Argentina. Parasitología Latinoamericana. 2005; 60:82–85. DOI: 10.4067/S0717-77122005000100015

[80] Motta CE, Rivero MR, De Angelo CD, Sbaffo AM, Tiranti KI. Risk and protective factors associated with gastrointestinal parasites of dogs from an urban area of Córdoba, Argentina. Turkish Journal of Veterinary and Animal Sciences. 2019; 43:846–851. https://journals.tubitak.gov.tr/veterinary/issues/vet-19-43-6/vet-43-6-17-1803-29.pdf

[81] Natalini B, Gennuso S, Beldomenico PM, Rigonatto T, Kowalewski MM. Parasitologic examination and associated risk factors of domestic dogs at the domestic-wildlife interface in the Iberá wetlands Ecoregion, Argentina. Veterinary Parasitology: Regional Studies and Reports. 2020; 20:100378. DOI: 10.1016/j.vprsr.2020.100378.

[82] Oku Y, Malgor R, Benavidez U, Carmona C, Kamiya H. Control program against hydatidosis and the decreased prevalence in Uruguay. International Congress Series. 2004; 1267:98–104. DOI: 10.1016/j.ics.2004.01.087.

[83] Olivares P, Valenzuela G, Tuemmers C, Parodi J. Description of parasites in faecal samples collected in plazas of downtown Temuco, Chile. Revista de Investigaciones Veterinarias del Perú. 2014; 25:406–413. https://doi.org/10.15381/rivep.v25i3.10119

[84] Opazo A, Barrientos C, Sanhueza AM, Urrutia N, Fernández I. Parasitic fauna in dogs (*Canis lupus familiaris*) of a rural sector in the central region of Chile. Revista de Investigaciones Veterinarias del Perú. 2019; 30:330–338. DOI: 10.15381/rivep.v30i1.15683.

[85] Oyarzún S, Cabello-Stom J, Moreira D, Hidalgo-Hermoso E. Estudio coproparasitario de perros (*Canis lupus familiaris*) que viven en co-ocurrencia con el zorro de darwin (*Lycalopex fulvipes*) en la cordillera de Nahuelbuta. Revista de Medicina Veterinaria e Investigación. 2019; 2:6–20. https://issuu.com/ussvet/docs/rev._med._vet._investig._n_4_2019

[86] Parra A, Orellana V, Rodríguez C, Valle M, Ricoy G, Santillán G. Evaluación de echinococcosis canina en la zona de alta montaña en la provincia de Tucumán, Argentina. Acta Bioquímica Clínica Latinoamericana. 2017; 51:133–137. https://www.redalyc.org/pdf/535/53550497016.pdf

[87] Pérez A, Costa MT, Cantoni G, Mancini S, Mercapide C, Herrero E, Volpe M, Araya D, Talmón G, Chiosso C, Vázquez G, del Carpio M, Santillan G, Larrieu E. Vigilancia epidemiológica de la equinococcosis quística en perros, establecimientos ganaderos y poblaciones humanas en la provincia de Río Negro. Medicina. 2006; 66:193–200. https://www.medicinabuenosaires.com/demo/revistas/vol66-06/3/VIGILANCIA%20EPIDEMIOLOGICA%

20DE%20LA%20EQUINOCOCCOSIS%20QUISTICA%20EN%20PERROS,.pdf

[88] Quilodrán-González D, Gädicke P, Junod T, Villaguala-Pacheco C, Landaeta-Aqueveque C. Risk factors associated with the presence of zoonotic gastrointestinal parasites in dogs of Cabrero District, Bio Bio Region, Chile. Chilean Journal of Agricultural and Animal Sciences. 2018; 34:118–125. DOI: 10.4067/S0719-38902018005000401

[89] Radman NE, Archelli SM, Burgos L, Domingo Fonrouge R, del Valle Guardis M. *Toxocara canis* en caninos. Prevalencia en la ciudad de La Plata. Acta Bioquímica Clínica Latinoamericana. 2006; 40:41–44. https://www.redalyc.org/pdf/535/53540107.pdf

[90] Rivero MR, Motta CE, Salas MM, Chiaretta A, Salomón OD. *Diphyllobothrium* sp. in *Canis familiaris* from the subtropical area of Argentina (Puerto Iguazú, Misiones). Revista Argentina de Microbiología. 2015; 47: 196–200. DOI: 10.1016/j.ram.2015.05.002.

[91] Rivero MR, De Angelo C, Nuñez P, Salas M, Motta CE, Chiaretta A, Salomón OD, Liang S. Environmental and socio-demographic individual, family and neighborhood factors associated with children intestinal parasitoses at Iguazú, in the subtropical northern border of Argentina. PLOS Neglected Tropical Diseases. 2017; 11: e0006098. DOI:10.1371/journal.pntd.0006098.

[92] Rodríguez F, Denegri G, Sardella N, Hollmann P. Relevamiento coproparasitológico de caninos ingresados al Centro Municipal de Zoonosis de Mar del Plata, Argentina. Revista Veterinaria. 2005; 16:9–12. https://revistas.unne.edu.ar/index.php/vet/article/view/1984/1731

[93] Roth D, Arbetman M, Flores V, Semenas L, Viozzi G. Diphyllobothriidea in the north area of the Andean Patagonia: Epidemiology in urban dogs, morphometrical and molecular identification, with comments on wild carnivores. Veterinary Parasitology: Regional Studies and Reports. 2018; 14:161–169. DOI: 10.1016/j.vprsr.2018.11.001.

[94] Rubel D, Zunino G, Santillán G, Wisnivesky C. Epidemiology of *Toxocara canis* in the dog population from two areas of different socioeconomic status, Greater Buenos Aires, Argentina. Veterinary Parasitology. 2003; 115:275–286. DOI: 10.1016/S0304-4017(03)00185-7.

[95] Rubel D, Wisnivesky C. Magnitude and distribution of canine fecal contamination and helminth eggs in two areas of different urban structure, Greater Buenos Aires, Argentina. Veterinary Parasitology. 2005; 133:339–347. DOI: 10.1016/S0304-4017(03)00185-7.

[96] Rubel D, Wisnivesky C. Contaminación fecal canina en plazas y veredas de Bs As. Medicina. 2010; 70: 355–363. https://www.medicinabuenosaires.com/demo/revistas/vol70-10/4/v70_n4_p355_363.pdf

[97] Rubel D, Carbajo A. Dogs in public spaces of Buenos Aires, Argentina: Exploring patterns of the abundance of dogs, the canine faecal contamination, the behaviour of people with dogs, and its relationships with demographic/economic variables. Preventive Veterinary Medicine. 2019; 170:104713. DOI: 10.1016/j.prevetmed.2019.104713.

[98] Sánchez P, Raso S, Torrecillas C, Mellado I, Ñancufil A, Oyarzo CM, Flores ME, Córdoba M, Minvielle MC, Basualdo JA. Contaminación biológica con heces caninas y parásitos intestinales en espacios públicos urbanos en dos ciudades de la provincia de Chubut, Patagonia, Argentina. Parasitología Latinoamericana. 2003; 58:

131–135. DOI: 10.4067/S0717-77122003000300008.

[99] Sánchez Thevenet P, Jensen O, Mellado I, Torrecillas C, Raso S, Flores ME, Minvielle MC, Basualdo JA. An eco-epidemiological study of contamination of soil with infective forms of intestinal parasites. European Journal of Epidemiology. 2003; 19:481–489. DOI: 10.1023/B:EJEP.0000027352.55755.58

[100] Semenas L, Flores V, Viozzi G, Vázquez G, Pérez A, Ritossa L. 2014. Helmintos zoonóticos en heces caninas de barrios de Bariloche (Río Negro, Patagonia, Argentina). Revista Argentina de Parasitología. 2014; 2:22–27. https://ri.conicet.gov.ar/handle/11336/12004

[101] Soriano SV, Pierangeli NB, Roccia I, Bergagna HFJ, Lazzarini LE, Celescinco A, Saiz MS, Kossman A, Contreras PA, Arias C, Basualdo JA. A wide diversity of zoonotic intestinal parasites infects urban and rural dogs in Neuquén, Patagonia, Argentina. Veterinary Parasitology. 2010; 167:81–85. DOI: 10.1016/j.vetpar.2009.09.048.

[102] Souto MG, Sánchez Thevenet P, Basualdo Farjat J. Evaluation of the presence of *Echinococcus granulosus sensu lato* in the environment and in hosts in a region endemic for hydatidosis in the province of Chubut (Argentina). Veterinary Parasitology. 2016; 6:42–46. DOI: 10.1016/j.vprsr.2016.09.005.

[103] Taranto NJ, Passamonte L, Marinconz R, De Marzi MC, Cajal SP, Malchiodi EL. Parasitosis zoonóticas transmitidas por perros en el chaco salteño, Salta, Argentina. Medicina. 2000; 60:217–220. http://www.medicinabuenosaires.com/revistas/vol60-00/2/parasitosis.htm.

[104] Torres P, Villalobos L, Woelfl S, Puga S. Identification of the copepod intermediate host of the introduced broad fish tapeworm, *Diphyllobothrium latum*, in southern Chile. Journal of Parasitology. 2004; 90:1190–1193. DOI: 10.1645/ge-3332rn.

[105] Vargas C, Torres P, Jercic MI, Lobos M, Oyarce A, Juan Carlos M, Ayala S. Frequency of anti-*Toxocara* spp. antibodies in individuals attended by the Centro de Salud Familiar and environmental contamination with *Toxocara canis* eggs in dog feces, in the coastal Niebla town, Chile. Revista do Instituto de Medicina Tropical de São Paulo. 2016; 58:62. DOI: 10.1590/S1678-9946201658062

[106] Winter M, Perera N, Marigual G, Corominas MJ, Mora M, Lecertua A, Ávila A, Arezo M. Enteroparásitos en heces caninas de la costanera pública de Viedma (Río Negro, Patagonia Argentina). Revista Argentina de Parasitología. 2018; 7:23–28. http://www.revargparasitologia.com.ar/pdf/RevArgParasitol_Vol7_Winter.pdf

[107] Zonta ML, Cociancic P, Oyhenart EE, Navone GT. Intestinal parasitosis, undernutrition and socio-environmental factors in schoolchildren from Clorinda Formosa, Argentina. Revista de Salud Pública. 2019; 21:224–231. DOI: 10.15446/rsap.V21n2.73692.

[108] Zunino MG, De Francesco MV, Kuruc JA, Schweigmann N, Wisnivesky MC, Jensen O. Contaminación por helmintos en espacios públicos de la provincia de Chubut, Argentina. Boletín chileno de parasitología. 2000; 55. DOI: 10.4067/S0365-94022000000300008.

[109] Haddaway NR, Bethel A, Dicks LV, Koricheva J, Macura B, Petrokofsky G, Pullin AS, Savilaakso S, Stewart GB. Eight problems with literature reviews and how to fix them. Nature Ecology & Evolution. 2020; 4:1582–1589. DOI: 10.1038/s41559-020-01295-x

[110] Savadelis MD, Evans CC, Mabry KH, LeFavi LN, Klink BD, Simson C, Moorhead AR. Canine gastrointestinal

nematode transmission potential in municipal dog parks in the southeast United States. Veterinary Parasitology: Regional Studies and Reports. 2019; 18: 100324. DOI: 10.1016/j.vprsr.2019.100324

[111] Hotez PJ. Human Parasitology and Parasitic Diseases: Heading Towards 2050. Advances in Parasitology. 2018; 100:29–38. DOI: 10.1016/bs.apar.2018.03.002

[112] Traversa, D. Are we paying too much attention to cardio-pulmonary nematodes and neglecting old-fashioned worms like *Trichuris vulpis*? Parasites Vectors. 2011; 4:32. DOI: 10.1186/1756-3305-4-32

[113] Felsmann M, Michalski M, Felsmann M, Sokół R, Szarek J, Strzyżewska-Worotyńska E. Invasive forms of canine endoparasites as a potential threat to public health - A review and own studies. Annals of Agricultural and Environmental Medicine. 2017; 24:245–249. DOI: 10.5604/12321966.1235019

[114] Rostami A, Riahi SM, Hofmann A, Ma G, Wang T, Behniafar H, Taghipour A, Fakhri Y, Spotin A, Chang BCH, Macpherson CNL, Hotez PJ, Gasser RB. Global prevalence of *Toxocara* infection in dogs. Advances in Parasitology. 2020; 109:561–583. DOI: 10.1016/bs.apar.2020.01.017

[115] Larrieu E, Gavidia CM, Lightowlers MW. Control of cystic echinococcosis: Background and prospects. Zoonoses and Public Health. 2019; 66:889–899. DOI:10.1111/zph.12649

[116] Larrieu E, Frider B, del Carpio M, Salvitti JC, Mercapide C, Pereyra R, Costa M, Odriozola M, Pérez A, Cantoni G, Sustercic J. Portadores asintomáticos de hidatidosis: epidemiología, diagnóstico y tratamiento. Revista Panamericana de Salud Pública. 2000; 8: 250–256. https://www.scielosp.org/pdf/rpsp/v8n4/3551.pdf

[117] Martínez PG. Hidatidosis humana: antecedentes generales y situación epidemiológica en Chile, 2001–2009. Revista Chilena de Infectología. 2011; 28:585–591. DOI: 10.4067/S0716-10182011000700013

[118] Vera GM, Venturelli FM, Ramírez JT, Venturelli AL. Hidatidosis humana. Cuadernos de Cirugía. 2003; 17:88–94. http://revistas.uach.cl/pdf/cuadcir/v17n1/art14.pdf

[119] López-Osorio S, Penagos-Tabares F, Chaparro-Gutiérrez JJ. Prevalence of *Toxocara* spp. in dogs and cats in South America (excluding Brazil). Advances in Parasitology. 2020; 109:743–778. DOI: org/10.1016/bs.apar.2020.01.029

[120] Eslahi AV, Badri M, Khorshidi A, Majidiani H, Hooshmand E, Hosseini H, Taghipour A, Foroutan M, Pestehchian N, Firoozeh F, Riahi SM, Zibaei M. Prevalence of *Toxocara* and *Toxascaris* infection among human and animals in Iran with meta-analysis approach. BMC Infectious Disease. 2020; 20:20. DOI: 10.1186/s12879-020-4759-8

[121] Fillaux J, Santillán G, Magnaval JF, Jensen O, Larrieu E, Sobrino-Becaria CD. Epidemiology of toxocariasis in a steppe environment: The Patagonia study. American Journal of Tropical Medicine and Hygiene. 2007; 76:1144–1147. DOI: 10.4269/ajtmh.2007.76.1144

[122] Martín UO, Demonte MA, Contini L, Giraldez E, Mendicino D, Del Barco M. Toxocariosis in different vulnerable groups of children in Argentina. Salud (i)Ciencia. 2014; 20:592–597. https://www.siicsalud.com/des/expertoimpreso.php/120904

[123] Godoy P, Mauvezin J, Basmadjián Y, Sayagués B, Giachetto G. Toxocariasis: manifestaciones clínicas y de laboratorio en niños asistidos en un

prestador integral de salud privado de Montevideo, Uruguay (2014–2018). Revista Médica del Uruguay. 2020; 36: 6–11. DOI: 10.29193/rmu.36.1.1

[124] Plascencia Gómez A, Proy H, Eljure N, Atoche Dieguez C, Calderón Rocher C, Bonifaz A. *Larva migrans* cutánea relacionada con *Ancylostomas*. Dermatología Revista Mexicana. 2013; 57:454–460. https://www.medigraphic.com/pdfs/derrevmex/rmd-2013/rmd136g.pdf

[125] Bava J, González LG, Seley CM, López GP, Troncoso A. A case report of cutaneous *larva migrans* in Argentina. Asian Pacific Journal of Tropical Biomedicine. 2011; 1:81–82. DOI: 10.1016/S2221-1691(11)60074-9

[126] González FCG, Galilea ONM, Pizarro CK. Autochthonous cutaneous *larva migrans* in Chile. A case report. Revista Chilena de Pediatría. 2015; 86: 426–9. DOI: 10.1016/j.rchipe.2015.07.018

[127] Kolbach M, Gamé A, Araya G. *Larva migrans* cutánea: reporte de un caso. Revista Chilena de Dermatología. 2012; 28:190. https://www.sochiderm.org/web/revista/28_2/17.pdf

[128] Beknazarova M, Barratt JLN, Bradbury RS, Lane M, Whiley H, Ross K. Detection of classic and cryptic *Strongyloides* genotypes by deep amplicon sequencing: A preliminary survey of dog and human specimens collected from remote Australian communities. PLOS Neglected Tropical Diseases. 2019; 13:e0007241. DOI: 10.1371/journal.pntd.0007241

[129] Rivero MR, De Angelo C, Núñez P, Salas M, Liang, S. Intestinal parasitism and nutritional status among indigenous children from the Argentinian Atlantic Forest: Determinants of enteroparasites infections in minority populations. Acta Tropica. 2018; 187:248–256. DOI: 10.1016/j.actatropica.2018.08.015

[130] Cimino RO, Fleitas P, Fernández M, Echazú A, Juarez M, Floridia-Yapur N, Cajal P, Seijo A, Abril M, Weinberg D, Piorno P, Caro N, Vargas P, Gil J, Crudo F, Krolewiecki A. Seroprevalence of the *Strongyloides stercoralis* infection in humans from Yungas Rainforest and Gran Chaco Region from Argentina and Bolivia. Pathogens. 2020; 9:394. DOI: 10.3390/pathogens9050394

[131] Mercado R, Jercic MI, Alcayaga S, de Paula FM, Ueta MT, Costa-Cruz JM. Seroepidemiological aspects of human *Strongyloides stercoralis* infections in Chile. Revista do Instituto de Medicina Tropical de São Paulo. 2007; 49:247–249. DOI: 10.1590/S0036-46652007000400010

Chapter 4

Incrimination of Dog Vector of Cystic Echinococcosis and Impact of the Appropriate Dogs' Treatment

Fatima Ezzahra Amarir, Abdelmohcine Aimrane and Abdelkbir Rhalem

Abstract

Dogs are involved in the transmission of several parasitic zoonosis. Among these, hydatidosis is very endemic in many countries of the world. Dog populations are very variable from one region to another, which increases the infestation risks across human populations especially in the developing countries such as in Morocco. Moreover, the risk of exposure is higher in dogs with access to rural slaughterhouses than in owned dogs. As for preventive measures, this calls for effective implementation of the appropriate dogs' treatment against hydatidosis. Thus, the following chapter updates the most relevant information on the impact of hydatidosis upon human populations and livestock animals, as to stretch understanding on the vector contribution of dogs.

Keywords: Dogs, Morocco, hydatidosis, zoonotic diseases, echinococcus granulosus

1. Introduction

Dog-borne zoonotic diseases include all the infectious diseases targeting dogs that can be transmitted to humans. Though they present major zoonosis causing a heavy burden upon the human population worldwide, these diseases are mostly neglected as few insufficient scientific research efforts are realized to face it (WHO, 2007b). Several examples of zoonosis are present with high prevalence up to time, such as bovine tuberculosis, brucellosis, cystic echinococcosis, visceral leishmaniosis, rabies [1], especially in the developing and North African countries in the poorest and most marginalized regions, in rural areas. Some of these diseases share the same definitive host represented by the dog. Especially hydatid cyst disease, which is endemic to hyper-endemic in agricultural countries in Europe, North, East and South Africa, South and North America, the Middle East and Asia [2–6]. Morocco is one of these highly endemic countries [7, 8].

The hydatid cyst, or *Cystic echinococcosis*, is caused by a small tapeworm parasite of canids, *Echinococcus granulosus*, which is then transmitted to humans via dogs. It has been reported in numerous reports that the incidence of the disease has increased in various parts of the world [9]. In Morocco for example, the annual incidence is 5.2 cases/100,000 inhabitants [10]. For this reason, several studies have been carried out in the world and in Morocco to evaluate chemoprevention in dogs in order to truncate the parasite cycle and reduce the incidence of the disease.

Hence, the purpose of this chapter is to compile recent data regarding the identification of the main source of infestation in dogs and the determination of the prevalence of infestation. As the canine population varies considerably from one region to another, like everywhere else in the world, in Morocco there are approximately 18,000 owned dogs and 3,000 stray dogs in endemic areas. A high prevalence of *E. granulosus* infestation in dogs has been recorded in these regions, estimated at 3.6% in Oulmes, 19.6% in Sidi Kacem and 23.7% in the Middle Atlas. This makes it possible to put the spotlight on the dog as the main reservoir and vector of this disease. Therefore, the risk of exposure is higher in stray dogs with access to rural slaughterhouses than in owned dogs, which is complicated to control. Hence, it is urgent the need for a very appropriate and regular chemoprevention program in dogs [7, 11].

2. Dogs vector of the major zoonotic diseases: an overview on *Echinococcus granulosus*

The dog has a high importance in the social life of the human population. These multiple and diverse functions make it an indispensable domestic animal, particularly for households in rural areas where the relationship between dogs and humans is very close. Unfortunately, the risk of transmission of pathogenic agents from dogs to other animals, mainly mammals, is an issue of major concern. Dogs indeed can act as reservoirs of pathogens as they may transmit *Leishmania spp*. (leishmaniasis), *Leptospira interrogans* (leptospirosis), *Toxoplasma gondii* (toxoplasmosis), *Neospora caninum* (neosporosis), *Dirofilaria immitis* (dirofilaria/heartworm disease), *Brucella canis* (brucellosis), *Sarcoptes scabiei* (scabies), *Echinococcus spp.* (echinococcosis), *Rickettsia rickettsii* (Brazilian spotted fever). Various canine viruses (e.g. distemper virus, adenovirus, coronavirus, herpes virus, parvovirus), rabies virus, among other pathogens for both humans and wildlife [12, 13]. Especially, *Echinococcus granulosus,* is up to now one of the zoonosis with a considerable endemic situation upon human populations and livestock animals.

Echinococcosis, the *Echinococcus granulosus* induced disease, is asymptomatic in dogs. Even with a high parasite load (from 1500 to 6000 worms per dog), this parasitosis may progress unperceived with no clinical signs. Moreover, due to the small size of the eliminated segments, no external signs can be seen. Nevertheless, anal pruritus can be induced following the penetration of gravid segments into the anal glands [14]. As the eggs are not visible by the human eye, there are no external signs of the infestation, explaining the danger of this parasite, which can easily spread and contaminate the environment, especially when dogs move from one place to another.

3. Impact of hydatidosis disease upon human's population and livestock animals production

Eggs of the *E. granulosus* parasite are disseminated in the environment by dogs. Thus, they are transmitted to a wide range of intermediate hosts, including sheep and humans, causing an infestation with the hydatid cyst (larval stage of the parasite) [7].

The abundance of stray dogs and slaughter practices that allow dogs' access to condemned offal, particularly in rural areas, contribute to the persistence of hydatidosis. Hydatidosis is a serious public health problem and has a significant socio-economic impact. The *Echinococcus granulosus* infestation is a major financial burden derived from human health costs and losses in livestock production. The economic burden of cystic echinococcosis on the global livestock industry has been estimated at over $2 billion per year. Despite the substantial socio-economic impact, hydatidosis is still a neglected zoonosis [15].

In humans, hydatid cyst is the cause of significant morbidity and mortality worldwide and is responsible for a significant economic loss in the public health sector [16, 17]. Hydatid cyst has several consequences, including the direct costs of diagnosis, hospitalization, surgical treatment, post-surgical care, for the patient and family members, without forgetting the indirect losses of mortality, pain and social consequences of lost working days and the cessation of agricultural activities by those affected or at risk [16–18]. People with hydatid cysts never restore a perfect health condition even after they have recovered [4].

At the livestock animal level, it involves losses in production, and their importance varies according to the breed and type of production concerned [19]:

- Organs not usable and seized at the slaughterhouse, especially liver and lung;
- Cost of destruction of infected viscera and dead animals;
- Possible restriction on the export of animals and their products;
- Parasitic hydatid cachexia associated with poly-parasitism in animals, which is a reason for reforming adult sheep whose productive life is reduced;
- Brutal mortality following the rupture of a hydatid cyst.

In sheep farming, it is estimated that 7–10% of milk losses, 5–20% of meat or whole carcass weight losses, and 10–40% of wool losses occurred (**Table 1**) [18]. In 1980, an assessment carried out in Italy [6] showed a 10% reduction in the commercial value of an infected sheep, a percentage which takes into account the cost of destroying viscera. It should be noted that the economic impact of infected viscera depends on the country's regulations and the number of animals slaughtered under veterinary control, as well as the cost of the equipment used [19]. According

Parameter	Reduction rate (%)	Reference
Cattle		[4]
Meat	2,5–10	
Milk	2,5–5	
Fertility	9,9 – 12,1	
Sheep		[4, 18]
Meat	5–10	
Wool	10–40	
Fertility	9,9 – 12,1	
Goat		[18]
Meat	5–20	
Fertility	9,9 – 12,1	
Camelin		[20]
Meat	2,5–10	

Table 1.
Reduction rate of animal products caused by hydatidosis [18].

to a recent study by Saadi et al., the economic impact of hydatidosis on animal production in Morocco is very significant [17].

4. Incrimination of dogs in transmissions of hydatidosis

4.1 Dogs infestation

In canids, particularly in dogs, infestation occurs by ingestion of intermediate host organs harboring the parasite at the larval stage (hydatid cyst). The protoscolexes released from the hydatid cyst grow into adult worms and live in the small intestine, particularly in the duodenum. The eggs are eliminated in the external environment by detaching the last proglottis from the mature worm and excreting it in the feces. In passage, some proglottis, which have been ruptured, release eggs at the marginal part of the anus. Anal pruritus provokes a licking reflex in the dog, which allows the dog to recover numerous eggs that will be found in the lingual papillae and the oral cavity and then, by licking, in the dog's pelage.

4.2 Relationship between dogs, human and livestock animals infestation

In Morocco, current evidence indicates that the transmission cycle of *E. granulosus* is mainly based on a domestic cycle involving dogs and livestock species (sheep, cattle, camels, goats and horses) [21]. According to a preliminary study carried out in the Middle Atlas, the prevalence of infestation in animals (all ages) is 29.82% in cattle (N = 102), 13.29% in sheep (N = 107) and 2.36% in goats (N = 16) (unpublished study). These regions of the Middle Atlas represent a hotspot of hydatid infestation with a prevalence of infestation of 91.7% in adult sheep (age > 4 years), and a prevalence of 1.9% in humans [22]. A large population of canids is present in these areas, which include owned dogs, stray dogs, jackals and foxes [23]. In 2019, the prevalence of *E. granulosus* reached 23–39% in owned dogs and 51–68% in stray dogs, while the risk of monthly incidence was 2–8% and 19–41% in owned and stray dogs, respectively [7]. In addition, the study conducted by Azlaf & Dakkak in various regions of Morocco revealed prevalence rates of 10.58% in sheep, 1.88% in goats, 22.98% in cattle, 12.03% in camels and 17.8% in horses [21]. The study conducted by El Berbri et al. in the region of Sidi Kacem revealed a prevalence of 42.9% in cattle, 11% in sheep and 1.5% in goats [24].

However, the abundance of dogs, especially stray dogs that eat infested offal in slaughterhouses and clandestine slaughter practices but also on farms that allow owned and sometimes stray dogs to feed on condemned offal, especially in rural areas, contribute to the persistence of hydatidosis. This represents a serious public health problem and has a significant socio-economic impact. The *Echinococcus granulosus* infestation is a major financial burden derived from human health costs and losses in livestock production.

Thus, one of the interesting models that reveals great relevance on the burden of *Echinococcus granulosus* on humans is the example of Morocco, of which the studies helped lot to extend the understanding of its various aspects. In the following section, we therefore put the focus on the most pertinent finding on the prevalence within the various categories of dog's populations in Morocco.

5. Categories of dogs' population in Morocco

In Morocco, the dog population is very diversified by the presence of different types of dogs: owned dogs and stray dogs or semi stray dogs. A study carried out in

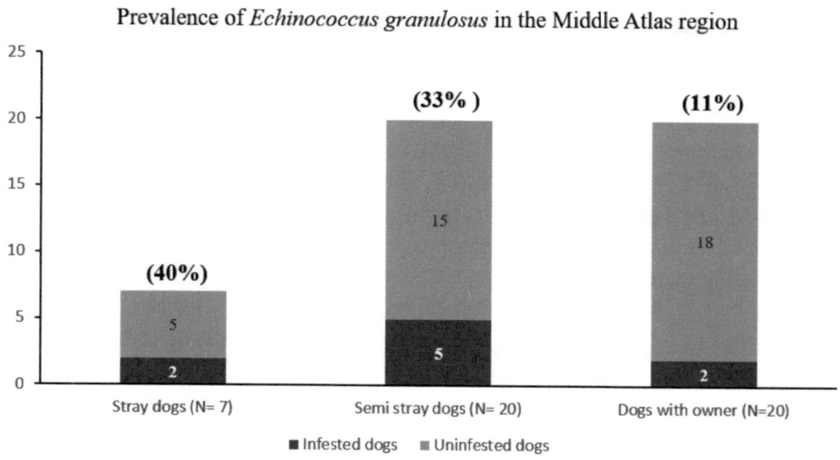

Figure 1.
Prevalence of Echinococcus granulosus *in three categories of dogs in the middle atlas region.*

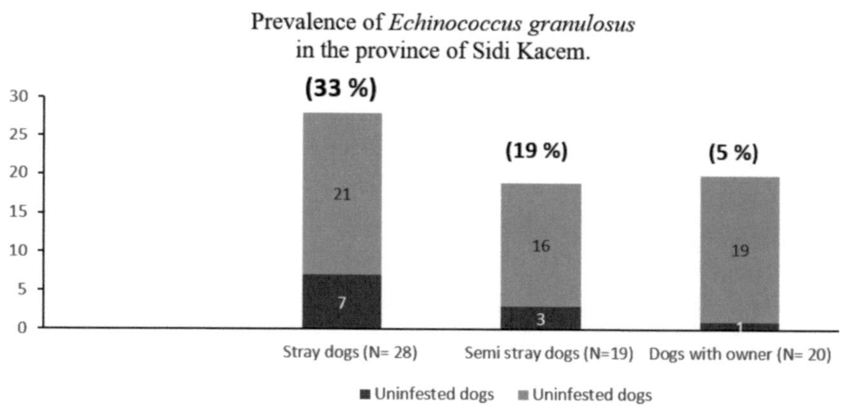

Figure 2.
Prevalence of Echinococcus granulosus *in three categories of dogs in the Sidi Kacem region.*

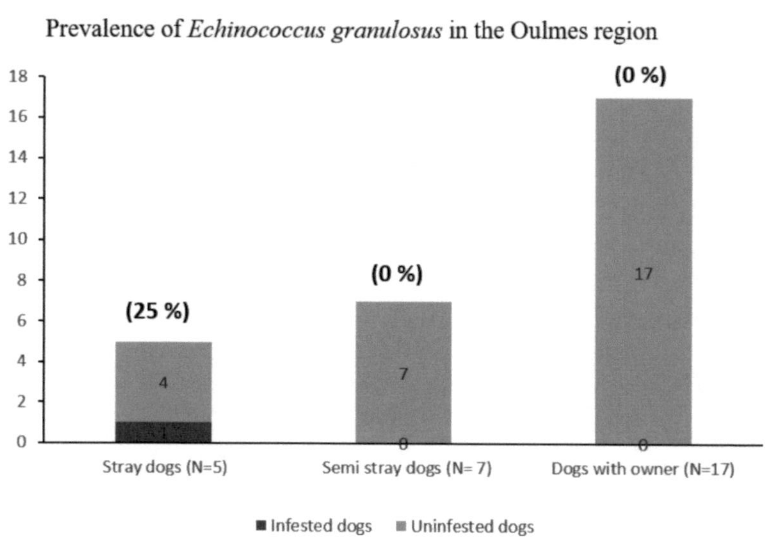

Figure 3.
Prevalence of Echinococcus granulosus *in three categories of dogs in the Oulmes region.*

three regions of Morocco revealed that stray dogs were the most infested category by *E. granulosus*, representing a prevalence of 40% in the Middle Atlas Mountains, 33% in Sidi Kacem and 25% in Oulmès, followed by semi-stray dogs and owned dogs (unpublished study) (**Figures 1–3**).

6. Diagnostics of echinococcosis in dogs

6.1 Clinical diagnosis

The identification of dogs infested with *E. granulosus* is extremely important for epidemiological studies and monitoring of this zoonosis in control programs [25]. Compared to other gastrointestinal infections of parasitic origin in dogs, echinococcosis is difficult to detect, even if the parasite load is high, premortem diagnosis appears to be very difficult. Recently, there has been considerable progress in research and development of immunological diagnosis for canine echinococcosis. The coproantigen test, in particular, can be considered to be of good sensitivity to reflect a common infestation and can therefore replace purging with arecoline (**Table 2**).

6.2 Parasitological diagnosis

The diagnosis of cystic echinococcosis in ante-mortem dogs can be made using several techniques:

6.2.1 Coproscopy

Coproscopy consists of looking for the eggs or proglottis of the adult worm in the feces of the final host. Eggs can be detected in fecal samples by the flotation technique or on the perianal skin by attaching a transparent adhesive paper to the skin and examining it under a magnifying glass under a microscope. However, this microscopic detection of *Echinococcus granulosus* eggs is not recommended because of the morphological nature of the egg which is similar in all taenia species. In addition, the removal of eggs is often irregular. However, proglottis from *Echinococcus granulosus* released spontaneously by dogs and detected on the surface of fecal samples can provide a good diagnosis of parasitosis in dogs [9].

6.2.2 Arecoline purging

Arecoline purging is the standard diagnostic method used for years in the detection of *Echinococcus granulosus* worms in dogs is purging the intestinal contents

Methods	Sensitivity %	Specificity %
Autopsy	> 90	100
Arecoline purging	50–70	100
Serology	35–70	> 90
The coproantigen test	75–80	> 95

Table 2.
Comparison of the sensitivity and specificity of different diagnostic methods for canine echinococcosis [12].

of the host with hydrobromide arecoline. Arecoline is a parasympathomimetic molecule which has an effect on the smooth muscles of the small intestine and at the same time paralyzes adult worms. The purgation removes the paralyzed worms with the feces, which must be inspected afterwards [9].

The advantage of this technique is the high specificity, which can reach 100%. On the other hand, the sensitivity does not even reach 50% if it is only used once. This test is contraindicated in pregnant dogs, older dogs and puppies. Arecoline should be administered orally at a dose of 4 mg/kg BW. This dose should be carefully calculated since severe undesirable secondary effects may occur [7].

6.2.3 Necropsy diagnosis: Sedimentation and counting

After cutting the intestine into several sections, these must be placed in metal trays, opened with scissors and finally immersed in physiological saline solution. The worms adhering to the mucous membrane are then counted using a magnifying glass or binocular microscope. The disadvantage of this method is that small worms can escape detection [9].

6.3 Immunological diagnosis by detection of coproantigens

This technique consists of searching for one of two types of antigens, either antigens extracted raw somatically from the worm or excretory-secretory antigens from the protoscolex in the feces of the host using double sandwich ELISA kits [26].

Positive ELISA results can be collected even in the prepatent period, starting on day 5 post-infestation. The values begin to decrease to negative values 2–4 days after the elimination of *Echinococcus granulosus* worms by treatment with praziquantel. The results of studies using this technique have shown that ELISA values correlate positively with the amount of worms present in the intestine, and that antigen levels are correlated with the amount of worms present in the intestine. This technique has been shown to be important with a sensitivity of 99% and a specificity of 97% [26].

Fecal samples can be taken directly from the ground or rectum and can be kept cold (−20°C) for up to 6 months. The test can be used for the identification of infected cases in control program, including pregnant dogs, older dogs and puppies. Three ELISA kits are commercially available today [27].

6.3.1 Serum antibody detection

Specific serum antibodies (IgG, IgA and IgE) can be detected in the serum of dogs infected with *Echinococcus granulosus* using antigenic preparations from the protoscolexes in ELISA kits. These antibodies can be detected 2–3 weeks post-infestation. One study suggests that eggs released in the small intestine of the final host, after proglottis apolysis, can penetrate the intestinal barrier and cause immunological stimulation in the host [27].

The ELISA kits available have low sensitivity and highly variable specificity. However, a new kit using a newly derived recombinant antigen from the protoscolex showed 100% specificity, but the sensitivity is not comparable to that of older kits. The use of ELISA kits for the detection of serum antibodies is still questionable because of their low sensitivity, the persistence of antibodies in serum after worm removal and the lack of correlation with infestation pressure [12, 26, 27].

If a seropositive test has been detected but the result is negative for the coproantigens, this is an indication of possible recent exposure [28].

6.4 Molecular biology diagnosis

Parasite DNA can be obtained from eggs, proglottis or worm cells and can be detected in feces after PCR amplification. However, no copro-PCR is currently available for the detection of all strains of *Echinococcus granulosus*; PCR primers for G1, G5 and combined G6/7 strains have been developed. This technique, due to its high cost, is only used for confirmation on positive samples in areas where the prevalence of cystic echinococcosis is low [29].

7. Treatment of infected dogs

Praziquantel is the only drug without significant undesirable effects known to be effective against *E. granulosus*. With a dose of 5 mg/kg, it can indeed achieve 100% efficacy [30, 31]. Because of its very broad therapeutic index, praziquantel is particularly suitable for cystic echinococcosis control programs [32]. Indeed, after its introduction in 1977, it was widely used in the majority of programs that undertook the control of the disease.

8. Prevention

Regular and accentuated treatment of stray dogs is necessary. However, regular treatment of owned dogs with Praziquantel should be an obligation in highly endemic areas, as treatment of dogs remains the most effective measure of prevention [7].

Vaccination of dogs with two recombinant proteins, EgA31 isolated from the adult worm and oncosphere and EgTrop isolated from protoscolex, is a promising approach to limiting the development of the *E. granulosus* worm in the dog's intestine [33].

9. Conclusion

In the case of zoonotic diseases, preventive veterinary treatments allow the protection of the public and animal health, but also the reduction of the risk of their transmission to humans, as is the case for cystic echinococcosis. To be effective, these treatments must be applied regularly. Thus, facility of access to them must be taken into account when developing the canine population management program. However, it should be noted that it is not only the dog that needs to be controlled, but also the intermediate host, and efforts should be made to eliminate the parasite or pathogen in general from the intermediate host that represents the main source of transmission to the dog to allow the pathogen to complete its life cycle and become infectious. Therefore the need for an integrated approach (action on the different hosts involved in the life cycle of the pathogen and the involvement of the socio-economic factor in control programs including stakeholders) to control these zoonosis is strongly advised.

DOI: http://dx.doi.org/10.5772/intechopen.95776

Author details

Fatima Ezzahra Amarir[1*], Abdelmohcine Aimrane[2] and Abdelkbir Rhalem[1]

1 Laboratory of Parasitology, Department of Pathology and Veterinary Public Health, Hassan II Agronomy and Veterinary Institute (IAV), Rabat, Morocco

2 Biochemistry Laboratory/Metabolic Platform, Cadi Ayyad University, Faculty of Medicine and Pharmacy, Marrakech, Morocco

*Address all correspondence to: amarir.f@gmail.com

References

[1] El Berbri I, Ducrotoy MJ, Petavy AF, Fassifihri O, Shaw AP, Bouslikhane M, et al. Knowledge, attitudes and practices with regard to the presence, transmission, impact, and control of cystic echinococcosis in Sidi Kacem Province, Morocco. Infect Dis Poverty. 2015;4: 1-12. doi:10.1186/s40249-015-0082-9

[2] Matossian RM, Rickard MD, Smyth JD. Hydatidosis: a global problem of increasing importance. Bull World Health Organ. 1977;55: 499-507.

[3] Wen H, Yang WG. Public health importance of cystic echinococcosis in China. Acta Trop. 1997;67: 133-145. doi:10.1016/S0001-706X(97)00051-X

[4] Torgerson PR, Heath DD. Transmission dynamics and control options for Echinococcus granulosus. Parasitology. 2003;127. doi:10.1017/S0031182003003810

[5] Budke CM, Campos-Ponce M, Qian W, Torgerson PR. A canine purgation study and risk factor analysis for echinococcosis in a high endemic region of the Tibetan plateau. Vet Parasitol. 2005;127: 43-49. doi:10.1016/j.vetpar.2004.08.024

[6] Dakkak A. Echinococcosis/hydatidosis: A severe threat in Mediterranean countries. Vet Parasitol. 2010;174: 2-11. doi:10.1016/j.vetpar.2010.08.009

[7] Amarir FE, Saadi A, Marcotty T, Rhalem A, Oukessou M, Sahibi H, et al. Cystic echinococcosis in three locations in the Middle Atlas, Morocco: Estimation of the infection rate in the dog reservoir. Vector-Borne Zoonotic Dis. 2020;20: 436-443. doi:10.1089/vbz.2019.2538

[8] Thys S, Sahibi H, Gabriël S, Rahali T, Lefèvre P, Rhalem A, et al. Community perception and knowledge of cystic echinococcosis in the High Atlas Mountains, Morocco. BMC Public Health. 2019;19: 1-15. doi:10.1186/s12889-018-6372-y

[9] Eckert J, Conraths FJ, Tackmann K. Echinococcosis: An emerging or re-emerging zoonosis? Int J Parasitol. 2000;30: 1283-1294. doi:10.1016/S0020-7519(00)00130-2

[10] Santé MDELA, Santé S De. d' Epidémiologie d' Intervention.

[11] Cabrera PA, Lloyd S, Haran G, Pineyro L, Parietti S, Gemmell MA, et al. Control of Echinococcus granulosus in Uruguay: Evaluation of different treatment intervals for dogs. Vet Parasitol. 2002;103: 333-340. doi:10.1016/S0304-4017(01)00603-3

[12] Craig PS, Gasser RB, Parada L, Cabrera P, Parietti S, Borgues C, et al. Diagnosis of canine echinococcosis: comparison of coproantigen and serum antibody tests with arecoline purgation in Uruguay. Vet Parasitol. 1995;56: 293-301. doi:10.1016/0304-4017(94)00680-B

[13] Campos-Bueno A, Lopez-Abente G, Andres-Cercadillo AM. Risk factors for Echinococcus granulosus infection: A case-control study. Am J Trop Med Hyg. 2000;62: 329-334. doi:10.4269/ajtmh.2000.62.329

[14] Veterinaire D. La Lutte Contre L ' Hydatidose En Sardaigne. 2009.

[15] Otero-Abad B, Torgerson PR. A Systematic Review of the Epidemiology of Echinococcosis in Domestic and Wild Animals. PLoS Negl Trop Dis. 2013;7. doi:10.1371/journal.pntd.0002249

[16] Craig PS, Larrieu E. Control of Cystic Echinococcosis/Hydatidosis: 1863-2002. Adv Parasitol. 2006;61: 443-508. doi:10.1016/S0065-308X(05)61011-1

[17] Saadi A, Amarir F, Filali H, Thys S, Rhalem A, Kirschvink N,

et al. The socio-economic burden of cystic echinococcosis in morocco: A combination of estimation method. PLoS Negl Trop Dis. 2020;14: 1-20. doi:10.1371/journal.pntd.0008410

[18] Battelli G. Echinococcosis: Costs, losses and social consequences of a neglected zoonosis. Vet Res Commun. 2009;33: 47-52. doi:10.1007/s11259-009-9247-y

[19] Battelli G. Socio-economic impact of cystic echinococcosis and of its control: Some data and considerations. Parassitologia. 2004;46: 359-362.

[20] Benner C, Carabin H, Sánchez-Serrano LP, Budke CM, Carmena D. Analysis of the economic impact of cystic echinococcosis in Spain. Bull World Health Organ. 2010;88: 49-57. doi:10.2471/BLT.09.066795

[21] Azlaf R, Dakkak A. Epidemiological study of the cystic echinococcosis in Morocco. Vet Parasitol. 2006;137: 83-93. doi:10.1016/j.vetpar.2006.01.003

[22] Chebli H, Laamrani El Idrissi A, Benazzouz M, Lmimouni BE, Nhammi H, Elabandouni M, et al. Human cystic echinococcosis in Morocco: Ultrasound screening in the Mid Atlas through an Italian-Moroccan partnership. PLoS Negl Trop Dis. 2017;11: 1-20. doi:10.1371/journal.pntd.0005384

[23] Maghnouj M. Quelques espèces de mammifères de la cédraie au Maroc : cas du singe magot. Forêt Méditerranéenne. 1999;20: 101-106.

[24] El Berbri I, Petavy AF, Umhang G, Bouslikhane M, Fassi Fihri O, Boué F, et al. Epidemiological Investigations on Cystic Echinococcosis in North-West (Sidi Kacem Province) Morocco: Infection in Ruminants. Adv Epidemiol. 2015;2015: 1-9. doi:10.1155/2015/104025

[25] Economides P, Christofi G, Gemmell MA. Control of Echinococcus granulosus in Cyprus and comparison with other island models. Vet Parasitol. 1998;79: 151-163. doi:10.1016/S0304-4017(98)00163-0

[26] Eckert J, Gemmell M, Meslin F-X, Z.S P. WHO/OIE Manual on Echinococcosis in Humans and Animals: A Public Health Problem of Global Concern. Veterinary Parasitology. 2002. doi:10.1016/s0304-4017(01)00631-8

[27] Zhang W, Li J, McManus DP. Concepts in Immunology and Diagnosis of Hydatid Disease. Clin Microbiol Rev. 2003;16: 18-36. doi:10.1128/CMR.16.1.18-36.2003

[28] Zeyhle E, Zhou P, Rogan M, Stringer R. Coproantigen detection for immunodiagnosis of echinococcosis and taeniasis in dogs and humans. Parasitology. 1992;104: 347-355. doi:10.1017/S0031182000061801

[29] Mathis A, Deplazes P. Copro-DNA tests for diagnosis of animal taeniid cestodes. Parasitol Int. 2006;55. doi:10.1016/j.parint.2005.11.012

[30] Farias LN, Malgor R, Cassaravilla C, Bragança C, De La Rue ML. Echinococcosis in Southern Brazil: Efforts toward implementation of a control program in Santana do Livramento, Rio Grande do Sul. Rev Inst Med Trop Sao Paulo. 2004;46: 153-156. doi:10.1590/s0036-46652004000300006

[31] Larrieu E, Zanini F. Critical analysis of cystic echinococcosis control programs and praziquantel use in South America, 1974-2010. Rev Panam Salud Publica/Pan Am J Public Heal. 2012;31: 81-87. doi:10.1590/S1020-49892012000100012

[32] Chemotherapie I, Ag B, Wuppertal D-. Parasitenkunde. 1978;179: 165-179.

[33] Lahmar PS. Travail sur Echinococcus en Tunisie. 2011; 4-6.

Chapter 5

Importance of Yeasts in Oral Canine Mucosa

Claudete Rodrigues Paula, Bianca Silva Navarro, Mário Mendes Bonci and Diana Costa Nascimento

Abstract

Dentistry science is a new specialty in veterinary medicine that has been growing in recent years, accompanied by the development of professionals who seek to improve the quality of life of pets. Cases related to problems in the oral cavity have gained significant importance in the medical clinic of professionals who treat small animals. Due to lack of professional knowledge or due to animal behavioral problems, such as aggressiveness, the anamnesis of the oral cavity is not performed most of the time, which ends up delaying the diagnosis of the pathology. In addition, an animal with a problem in the oral cavity may take years to show signs of the disease. In general, animals have an oral microbiota composed of various species of fungi, which, under specific conditions, can change from saprophytes to pathogens, compromising their health. Thus, the pre-knowledge of potentially pathogenic yeasts belonging to oral microbiota of dogs and their susceptibility profile compared to the main drugs used in antifungal therapy, is of fundamental importance as it ensures a clinical auxiliary support for the diagnosis and treatment of most diseases of the oral cavity.

Keywords: yeasts, oral cavity, dogs, antifungal, microbial resistance, fungi

1. Introduction

Fungi are eukaryotic, heterotrophic organisms, multinucleated like molds, or only with one nucleus, like yeasts. These organisms can be unicellular, or multicellular, which we call mycelium. Yeasts are unicellular and do not present, in general, morphological differences. The cells are rounded, ovoid or elongated, but some yeast under special conditions may have successive sprouts in a chain, which we call pseudomycelium [1].

The classification of fungi is based on morphological, reproductive and physiological characteristics. The taxonomy of fungi is still varied, but we can classify them in the Kingdom Fungi in the six phylas: Basidyomycota, Ascomycota, Glomeromycota, Chytridiomycota, Blastocladiomycota and Neocallimastigomycota [2, 3].

Approximately 200 out of a total of 100,000 species of yeast are considered pathogenic. Of these pathogenic species, 50 of them are regularly associated with mycoses. Yeasts are the ones that cause the greatest number of mycoses, both in man and in animals and we highlight the genera *Candida*, *Cryptococcus*, *Malassezia* and *Trichosporon* [3].

Figure 1.
Mixed breed dog.

These yeasts can be assexual (anascoporogenous), or sexual (ascoporogenous or basidioporogenous). In general, they are considered opportunists "waiting" for their "opportunity", that is, the drop in the immunity of man and animals, thus causing a case of ringworm.

Among domestic animals, the ones that have the closest proximity to people are dogs. *Canis lupus familiaris* is believed to have emerged approximately 130,000 years ago, from the domestication of the gray wolf. Crossbreeding and selection of characteristics gave rise to different breeds, including Poodle, Yorkshire, Terrier and Labrador Retriever, but mixed breed animals are prevalent in homes around the world (**Figure 1**) [4].

In addition to being mere companions in people's homes dogs have established themselves with essential functions such as security and hunting. These dogs have gained these and other noble functions and thus brought them even closer to human beings in places and situations that would otherwise be dispensed with. Today they also act as guides for the visually impaired, accompanying people to the hospital, monitoring blood glucose levels for diabetic people and even detecting pathogens in hospital environments [5].

These new functions, with consequently greater proximity between dogs and people, also result in a possible greater exchange of microorganisms between these beings, including yeasts. Among these fungi, the most present in the oral mucosa of dogs are the genera *Malassezia* and *Candida* and found less the genus *Cryptococcus* [6].

In the field of public health, these microorganisms have in common the ability to cause disease in both animals and people, and therefore this possible increase in the sharing of microbiota between these beings must be monitored by health specialists.

It is important to emphasize that the exchange of microorganisms occurs in both directions, and that the health of the animals must also be considered in these cases.

The vigilance of the clinical mycologist must be maintained for a better understanding of how future changes can become serious public health problems, especially for yeasts, as we have already seen in several situations.

2. Ecology and sources of yeast infection

Yeasts can be found in plants, soil, air, aquatic environment, in invertebrate and vertebrate animals, that is, in almost all ecosystems. These microorganisms can be in their symbiotic state, in mutualism, or in parasitism. In humans, several species can be part of their natural microbiota, in the gastrointestinal tract, in mucocutaneous tissues and skin. In man, a large part of yeast infections, especially of the genus *Candida*, are of endogenous origin and are linked to risk factors such as old age, prematurity, avitaminosis, antibiotic therapy, cancer, and other diseases that cause immunodepression of the host [7].

Extrinsic factors can also be important, such as the rupture of the natural barrier of the skin and mucous membranes, the use of invasive hospital material and contact with contaminated ecological niches. Direct transmission between people can occur in sexual relations [8].

In dogs, the main yeast found on the skin and mucocutaneous surfaces is *Malassezia pachydermatis*, which easily recovers in the folds of the skin and especially in the various parts of the ear. The prevalence of some types of yeasts in the oral mucosa of dogs is related to several habits, such as licking, sniffing and exploring environments.

The licking of the paws and other areas of the body explains the considerable presence of *Malassezia pachydermatis* in the oral cavity of dogs. Considered a saprophyte in the skin of dogs, this microorganism can cause dermatitis in several situations, and in these cases, there is also an increase in its presence in the oral mucosa. Other relevant yeasts of these animals belong to the genera *Candida*, *Rhodotorula* and *Trichosporon*, which are, in most cases, in balance with the dogs' organism [9].

It is also reported that *Cyniclomyces guttulatus*, present in the stomach, intestine and feces, which in situations of imbalance with the commensal microbiota, may be related to clinical conditions that affect the gastrointestinal tract [10].

The habit of sniffing the soil, in parks and gardens, hunting in forests and dens, favors the sharing of microorganisms among animals linked to these environments. The organic matter present in these places, mainly in the feces of birds and bats, favors colonization by fungi such as *Cryptococcus* spp. and *Histoplasma capsulatum*, which in situations favorable to microorganisms (host immunosuppression; high microbial inoculum load) can cause serious diseases [11].

Advances in veterinary hospital techniques, especially surgical procedures and hospitalizations, also brings new sources of infection for dogs. The ability of microorganisms of the genera *Candida* and *Malassezia* to form biofilms makes equipment such as specula, probes and other surgical materials possible sources of transmission of these microorganisms. For this reason, the correct asepsis and sterilization for handling this equipment is extremely important to avoid mycoses and severe cases of fungemia [12].

3. Predisposing and virulence factors to yeast infections

There are several yeasts that are of interest to the veterinarian, which can cause superficial, subcutaneous, mucosal lesions, and even granulomatous and systemic processes, and, in most cases, suspicion about the fungal etiology of cases is neglected, hence advanced and severe cases of mycosis in dogs are not uncommon [13].

The transition from the yeast stage to commensal to pathogenic will depend both on factors related to the agent's virulence, as well as on the host's own susceptibility [14].

The factors that can predispose humans and animals to a yeast infection are innumerable, resulting from alterations in the defense mechanisms or by compromising the anatomical barriers of protection of the organism [15–17].

Among these factors we can mention: stress; use of broad-spectrum antibiotics or prolonged antibiotic therapy; antineoplastic agents; neutropenia; immunosuppression; age (senility/puppy); inadequate environment (overcrowding); long-term use of corticosteroids; nutritional deficiencies; diets with a high concentration of carbohydrates; pH changes, vitamin A deficiency, trichomoniasis; presence of autoimmune diseases; changes in anatomical barriers due to trauma (maceration); aplastic anemia; hematological infections; periodontal diseases (**Figure 2**) and other concomitant diseases [17, 18].

Prolonged antibiotic therapy and a high concentration of carbohydrates in the diet can lead to the destruction or inhibition of the competitive bacterial microbiota, disrupting its balance with the host organism, thus allowing the accentuated growth of yeasts [15].

Probably due to the poor oral hygiene of dogs throughout their life and associated with the other predisposing factors already mentioned, senility is considered a significant condition for predisposition to periodontal disease. Animals older than 4 years, according to a study with stray dogs, are more likely to develop this disease, ranging from mild gingivitis to severe periodontitis (**Figure 3**).

Virulence factors attributed to microorganisms must also be taken into account, such as production of hydrolytic enzymes, proteases and phospholipases, adhesion, formation of germ tube and biofilms. These factors favor the invasive power and interfere with the host's metabolism. All these factors, from hosts and yeasts, can lead to superficial, or systemic, conditions. It is worth mentioning that the high concentration of viable cells of the microorganism in an ecological niche of the host is another factor that must be considered, as they may be part of the oral microbiota.

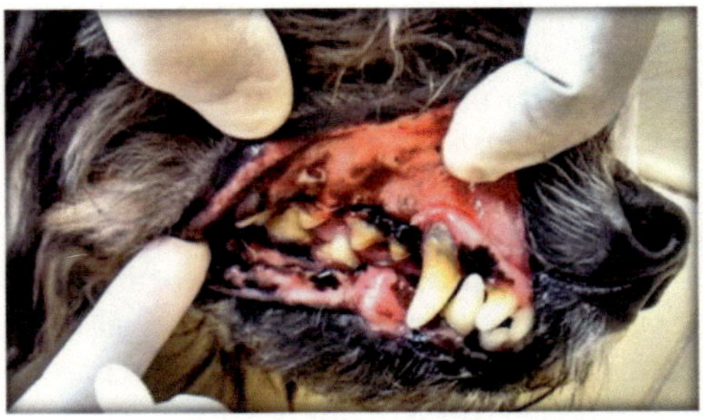

Figure 2.
Dog with periodontal disease.

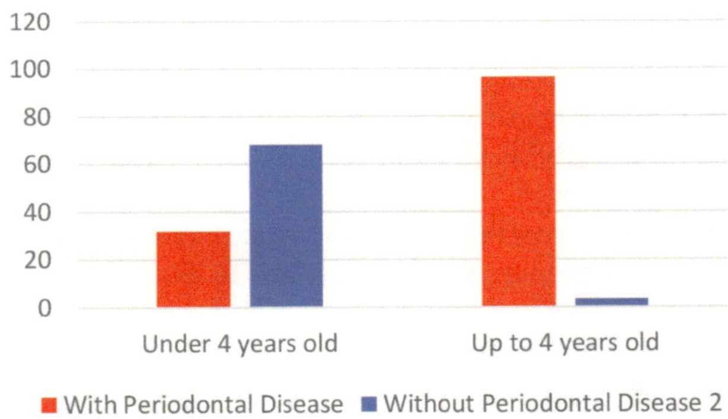

Figure 3.
Connection: Age x Presence of Periodontal Disease in dogs [19].

4. Importance of animals' oral health

Disorders of the oral cavity are of great importance in veterinary medicine due to their high prevalence in dogs and their serious consequences, which can even affect the systemic health of the animal [20]. Abnormalities, injuries or disorders of this organ can cause discomfort and pain, leading the animal to anorexia, due to lack of food, and adipsia, not water intake, predisposing it to conditions of decreased immunity and clinical complications [21].

In addition to this great discomfort and the involvement of other organs, the inflammatory response caused by diseases in the oral cavity can lead to the gingival tissue a progressive loss of tooth fixation to the alveolar bone and, consequently, the loosening and loss of this tooth [22].

The dentistry specialty in veterinary medicine has been evolving in recent years, gaining space in the curriculum of some colleges. Even though the food industry has undergone great advances in the production of diets aimed at improving oral health, the number of professionals who perform an adequate clinical examination is still not significant. In addition to this important factor, the lack of adequate provision of oral hygiene care is worrying [22, 23].

Among dogs over one year of age, 95% have some degree of the disease, and in the clinic, it is believed that 100% of adult animals have varying degrees of periodontal disease [23]. The most common signs associated with periodontal disease are halitosis, dental calculus, inflammation and gingival bleeding, anorexia and the consequent weight loss, ptyalism, difficulty in chewing and grinding food, mobility and migration of teeth, loss of alveolar bone, gingival retraction and behavioral changes [24].

Periodontal disease is, therefore, the most common disease affecting dogs of all breeds, formed from proliferative microorganisms, defense cells (leukocytes and macrophages), epithelial cells, bacterial polysaccharides and salivary glycoproteins, which over time become organize, occurring mineralization and formation of dental calculus [25]. It is believed that this clinical condition is usually caused by the formation of bacterial plaques, but the isolation of yeasts from the oral cavity of dogs with periodontal disease is frequent (**Figure 4**).

The greatest risk in periodontopathic is not only the loss of teeth or the development of local infections, but the possible systemic effects of the pathological agent in the bloodstream.

Thus, the oral health of dogs is extremely important and still needs a greater focus on microbiological research and awareness of those responsible, regarding food, the importance of oral hygiene, and the attention of the tutor and the veterinarian regarding the etiopathogenesis of diseases, such as yeasts.

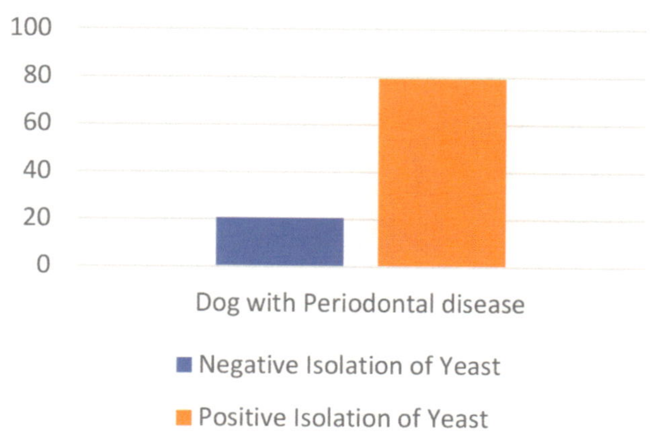

Figure 4.
Connection: Presence of Periodontal Disease x Positive isolation of yeast [19].

5. Main genera and species of yeasts isolated from the oral cavity of dogs and clinical signs

Just like in humans, dogs have a known range of yeasts in their oral mucosa that still requires more studies regarding colonization and pathogenicity. Despite its remarkable importance in the health of dogs, studies involving the isolation and correct identification of yeasts began to be developed in the 20th century [26].

This microbiota is not yet fully described, due to its great complexity and diversity. Fungal colonization of the oral cavity of dogs is associated with yeasts of the genera *Candida*, *Malassezia*, *Trichosporon* and *Rhodotorula*. Less frequently, we can isolate yeasts of the genus *Cryptococcus* [27].

In a recent study conducted with 50 mixed breed dogs, a yeast profile was found, composed of *Candida albicans* (39.5%), *C. parapsilosis* (18.6%), *C. zeylanoides* (13.9%), *C. krusei* (7%), *C. tropicalis* (4.7%), *Trichosporon* spp. (4.7%), *T. asahii* (4.7%), *C. guilliermondii* (2.3%), *T. mucoides* (2.3%) and *Malassezia pachydermatis* (2.3%). The genus *Candida* showed a high prevalence, making up a total of 82.2% of the isolated yeast profile. It is worth mentioning here the isolation of *Candida zeylanoides*, a rare species, even in humans, and thus, the oral mucosa of dogs can harbor a new "ecological niche" of this fungus species, which can also act as an opportunistic pathogen [9].

5.1 Genus *Candida* and Candidiasis

Currently, 317 species of this genus are recognized. Several of these species, more precisely 20, have a pathogenic potential and can thus cause infections in several species of animals, such as dogs [17]. The relationship with the host can be commensal, parasitic or saprophytic. It can also be found in the usual form of a yeast, or in the form of pseudohyphae. *Candida albicans* is the most common colonizer in cases of infections, with a predilection for mucous surfaces and mucocutaneous areas. Other species, such as *C. kefyr, C. lusitaniae, C. guilliermondii, C. tropicalis, C. krusei, C. famata, C. parapsilosis*, can be isolated from animals (**Figure 5**) [15].

There were only few cases found in the literature in small animals, however, reports of candidiasis in various animal species are also increasingly common, described in photos of pyoderma of the lip folds, disseminated and localized mucocutaneous dermatitis, urinary tract infections, gastrointestinal and reproductive system, ear infections, systemic and oral infections [27].

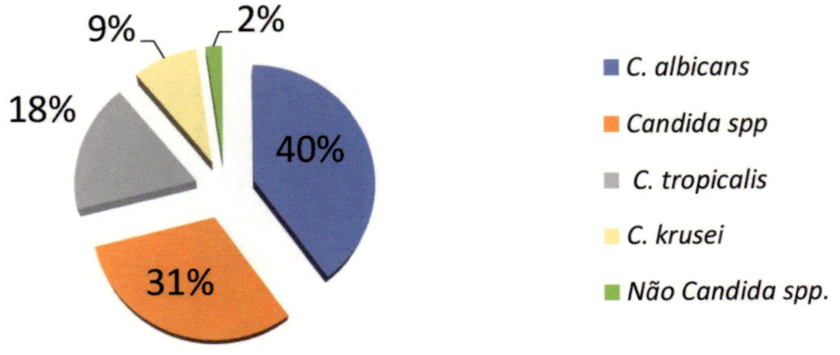

Figure 5.
Presumptive result of Candida *species isolated from the oral cavity of mixed breed dogs according to the CHROMAGAR Candida®.*

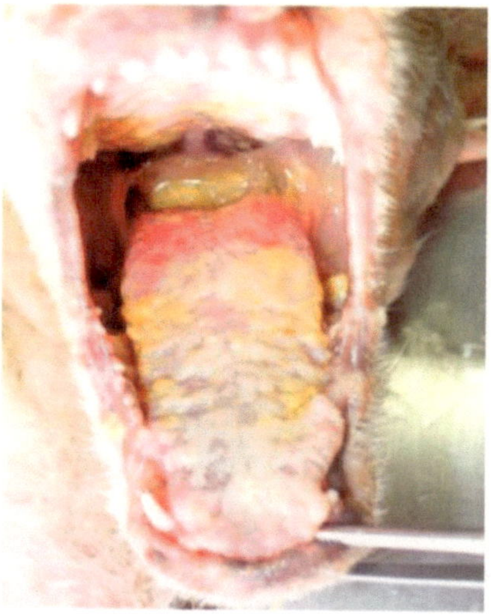

Figure 6.
Mixed breed dog with oral candidiasis (glossitis). Friable white and yellowish plates covering the tongue [28].

Candidiasis related to the digestive system of dogs, such as a clinical manifestation of glossitis, is characterized by the formation of pseudomembranous plaques, usually whitish in color, or yellowish beige. Once these plaques were removed, we noticed erythematous regions with the presence of ulcers (**Figure 6**) [28].

5.2 Genus *Trichosporon* and Trichosporonose

The genus *Trichosporon* has 37 species that inhabit different ecological niches, such as water, soil and body and mucous surfaces of humans and animals. They can cause superficial and deep infections, such as *Trichosporon asahii*, *T. mucoides*, *T. ovoides*, *T. inkin*, *T. asteroides* and *T. cutaneum* [29].

No cases of *Trichosporon* infections have been reported in the oral mucosa of dogs, however, several species have already been isolated as colonizers. Clinical cases of nasal granuloma in other animals, cystitis in cats, mastitis in cows and dermatitis in horses and monkeys have already been described (**Figure 7**) [30].

5.3 Genus *Malassezia* and Malasseziosis

The genus *Malassezia* has 15 species, mostly lipophilic yeasts, that can be part of the skin and mucous membranes of humans and dogs. They are opportunistic yeasts, and in certain circumstances, they can lead to clinical manifestations [31]. *Malassezia pachydermatis*, the most frequent in dogs, is not lipophilic and can grow in a culture medium common in mycology [19] (**Figure 8**).

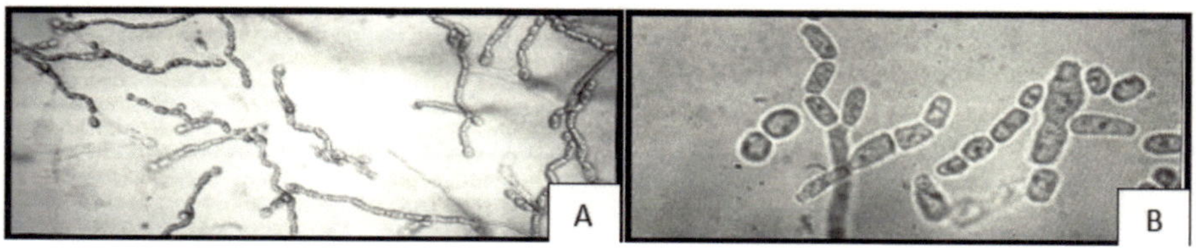

Figure 7.
A and B –Yeasts of Genus Trichosporon *showing oval and rectangular artrospores - (A) 160x and (B) 400x [15].*

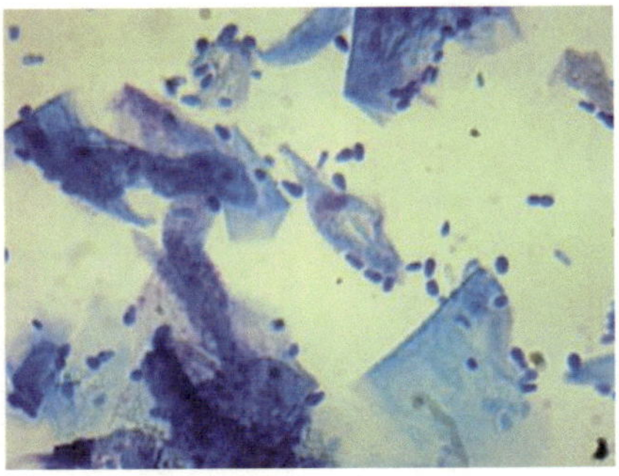

Figure 8.
Yeasts of the Genus Malassezia *spp. - single budding on wide base - Panotic, 1000x.*

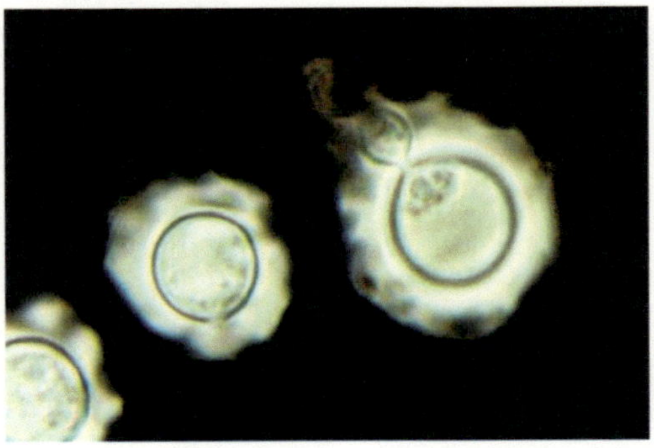

Figure 9.
Cryptococcus *spp. Encapsulated yeasts - Nigrosina, 1000x.*

Several clinical symptoms can be associated with *Malassezia* spp. and, particularly, in cases of otitis and dermatitis in dogs. Cases of otitis by *Malassezia pachydermatis* are frequency, but oral infection caused by this agent have not been described or has not yet been well studied [19].

5.4 Genus *Cryptococcus* and cryptococcosis

In the *Cryptococcus* genus, we found 38 species, with *Cryptococcus neoformans* and *C. gattii* being the most prominent in medical mycology in man and animals (**Figure 9**). The species can be found in different places in the environment, primarily in association with birds' droppings, mainly pigeons, but have an ecological association with trees too, such as eucalyptus [32].

In dogs, can enter the body through the lung causing pulmonary disease, and several clinical signs can be presented, such as skin lesions, nasal mucosa ("clown nose"), and can hit the central nervous system, for its neurotropic nature. These lesions in the nasal mucosa can extend into the oral cavity of the animals [28].

6. Collection of clinical material and identification of yeasts

6.1 Material collection

The collection of the oral cavity of dogs is performed with the aid of a sterile, alginate swab, moistened with sterile saline solution. The swab is introduced,

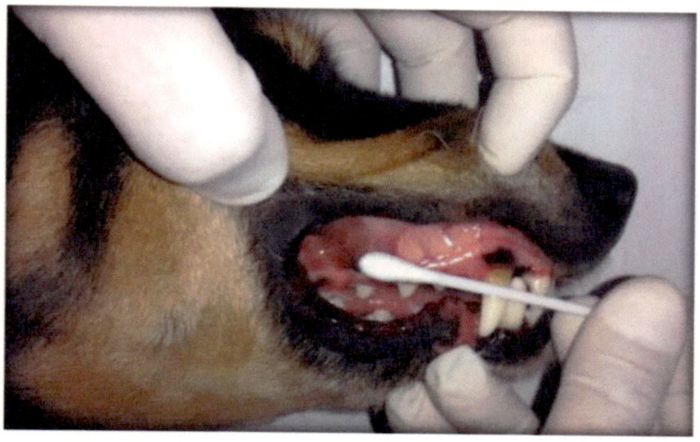

Figure 10.
Collection with sterile swab of the oral mucosa of a mixed breed dog - City of Campinas, São Paulo - Brazil.

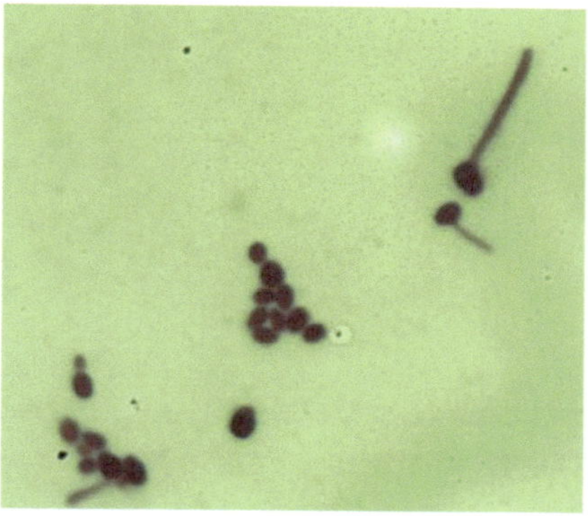

Figure 11.
Yeasts of Genus Candida *in an abdominal dog fluid sample - Fuchsin, 1000x.*

carefully, in the oral cavity, in circular movements, passing through the entire oral mucosa [19] (**Figure 10**).

After this procedure, the collected samples must be sent to the laboratory and sown in Petri dishescontaining basic mycology medium (Sabouraud dextrose agar), plus antibiotics (chloramphenicol - 0.05 g/L concentration). Incubation at 25°C for up to two weeks [19].

There are several procedures that can be used to identify yeasts. Direct examination (fresh), or with Gram stain is also highlighted (**Figure 11**).

6.2 Yeast identification

The identification of yeasts can be performed by means of macro and micromorphological, biochemical, proteome (MALDI-TOF) and molecular tests. In macromorphological characterization, we studied color, texture and edges (**Figure 12**).

In the more specific micro morphological identification, we must observe the characteristics of the cells (oval, round, unipolar bud, or multiple buds), pseudohyphae, hyphae and structures characteristic of *C. albicans*, such as chlamydoconidia (**Figure 13**).

The formation of a germ tube, another important characteristic of *C. albicans*, originates from blastoconidium when the yeast is sown in fetal bovine serum (**Figure 14**).

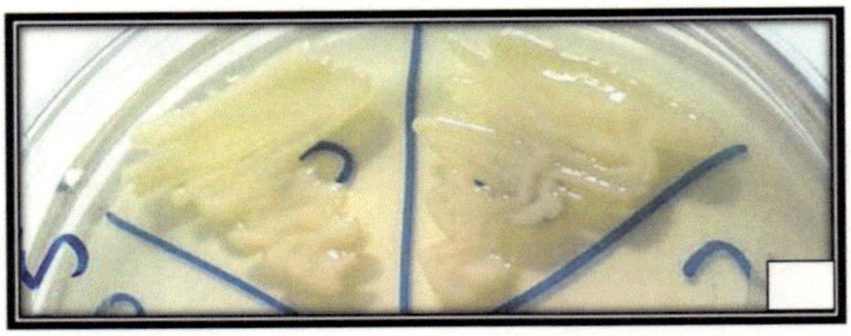

Figure 12.
Culture of yeasts on Sabouraud dextrose agar.

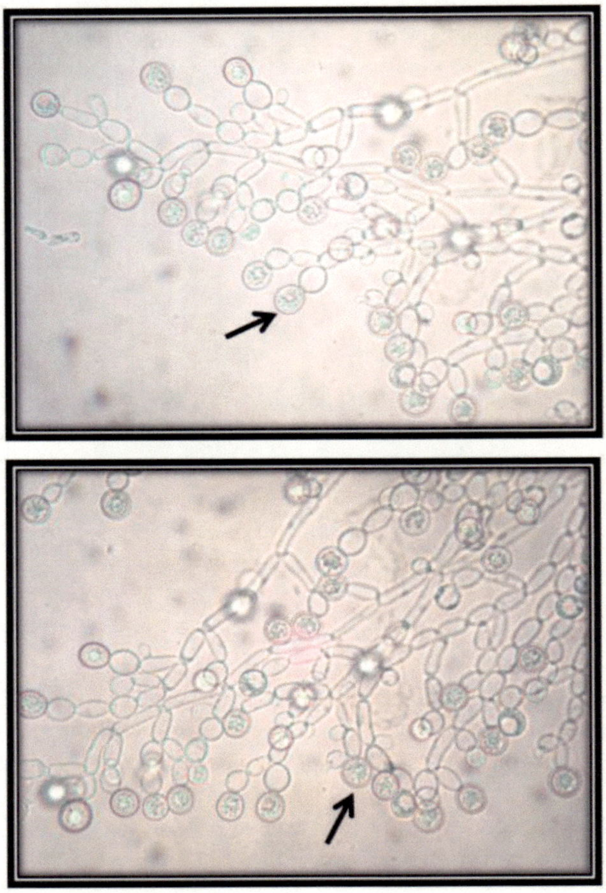

Figure 13.
Candida albicans *in culture broth. Globose, or elongated cells, pseudohyphae, hyphae, blastoconidia and characteristic chlamydoconidia, 1000x.*

Tests of assimilation of sources of nitrogen and carbohydrates can be performed (auxanogram), as well as fermentation tests (zymogram). The protocol followed for these methods is from the manual "The Yeasts: a taxonomic study" (volumes 1, 2 and 3). The MALDI-TOF technique is a mass spectrometry, which determines the protein profile (proteome) of the yeast under study. It is a fast technique (15–20), simple, excellent cost–benefit, however, there are limitations to the use of the laboratory routine, as the device is expensive and requires specialists to use it, as well as a robust base of standard strains.

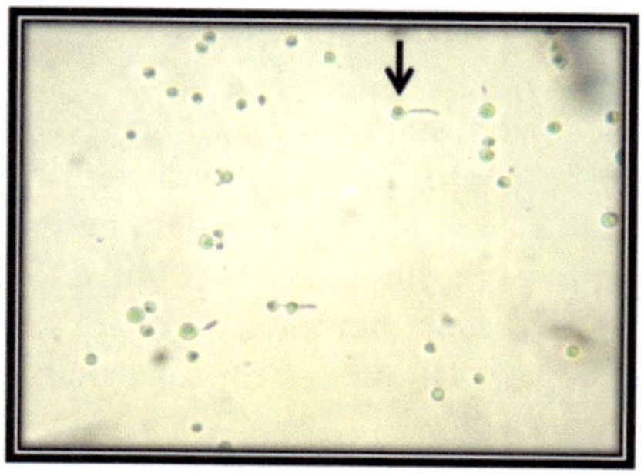

Figure 14.
Germ tube on bovine serum – Candida albicans*, 400x.*

For the identification of yeasts, we also count on molecular biology techniques, which are sensitive and specific. For the differentiation, for example, of *C. albicans* and *C. dubliniensis* is the most accurate technique. There are several methods such as PCR (Polymerase chain reaction), RFLP (Restriction fragment length polymorphism) and RAPD (Random amplified polymorphic DNA). One of the most used is PCR, which detects minimal amounts of DNA, or RNA. But not all laboratories can use these methods, due to the higher costs and the needs of specialized laboratories [33].

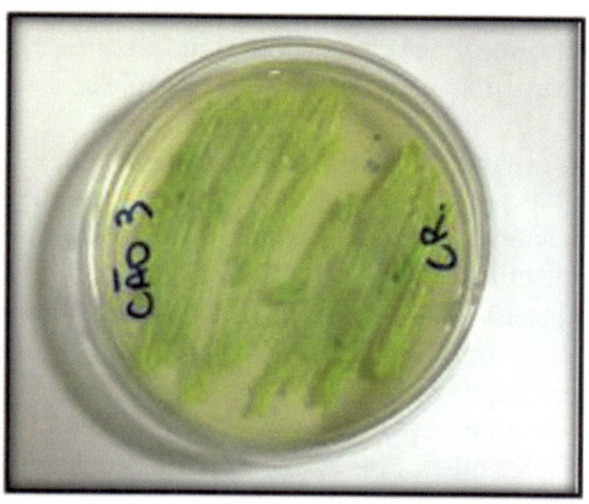

Figure 15.
Yeasts of genus Candida on CHROMAGAR Candida® - Candida albicans *with green color.*

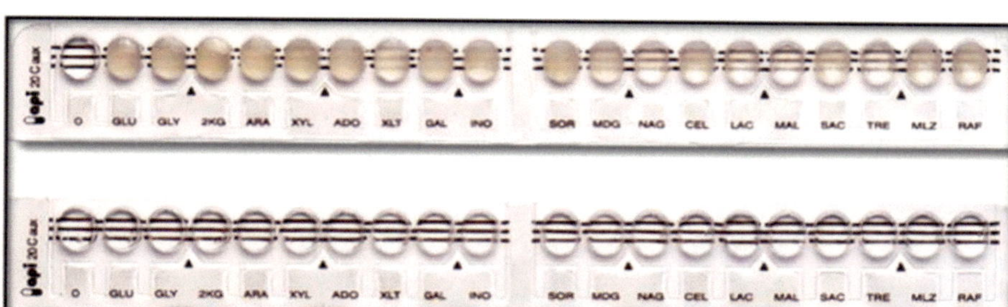

Figure 16.
API20CAUX method (bioMérieux®) –profile of carbohydrates assimilation.

6.3 Chromogenic medium: CHROMagarCandida®

Sowing in chromogenic media, such as CHROMagarCandida®, can provide presumptive identification according to the color developed by the yeast. In this medium, the specie *Candida albicans* develops a light green color; *C. tropicalis* it is blue/green and *C. krusei* light pink, for example (**Figure 15**).

In addition to these identification methods, there are several automated and manual systems that facilitate the laboratory routine, such as Vitek and API20C, as examples (**Figure 16**).

7. Epidemiological markers

As we have already pointed out, yeasts (especially those of the genus *Candida*) have emerged as important pathogens in humans and animals and the interrelation between both is of great relevance, gaining prominence today.

Several markers can be used to detect new yeast species as well as their genotype. Based on the data, we can determine the presence of these microorganisms, the same species/genotype, in one or more anatomical areas of the host, as well as of different ecological niches.

Confirming the colonization/infection area has often been an arduous task. Therefore, the use of these markers shows to be of great importance for the epidemiological study of yeasts.

Among the phenotypic markers, we can highlight those based on colony morphology (morphotyping), enzyme production (enzyme typing), sensitivity to "Killer" toxins and antifungal agents. They are simple and easy to perform techniques.

Genotypic markers are more sophisticated and safer; however, they require more elaborate techniques. The technique is based on short sequential repetition of bases throughout the yeast genome and its reading is performed on a specific sequencing apparatus. The patterns of the visualized DNA bands function as true "fingerprints" of the microorganism, leading us to the recognition of the colonization/infection area of the host. This technique can be used both for use in clinical isolates and for environmental samples (**Figure 17**).

With increasingly interconnected ties between man and his dog, the use of these markers is a valuable technique for detecting epidemiological transmission between

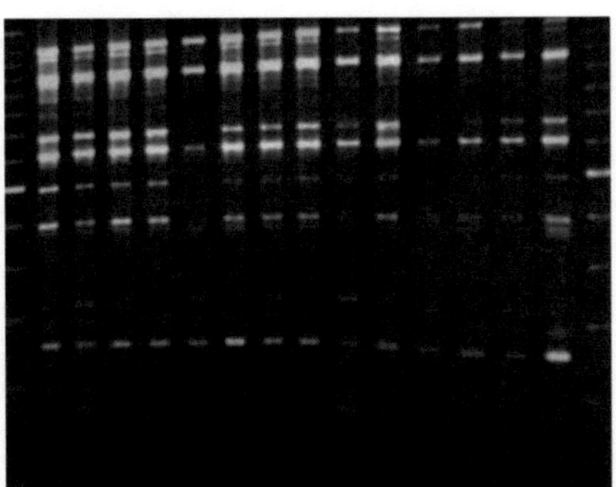

Figure 17.
Chromosomal bands of yeasts obtained by electrophoresis pulsed field - PFGE.

these species and a facilitator for taking therapeutic actions based on the microbiological analysis of the agent's transmitter [33].

8. Antifungals, sensitivity tests and treatment

Currently, yeast mycoses have increased substantially, and it can be considered an important public health problem, especially in systemic clinical conditions and hospital infections. The antifungal drugs used in human and veterinary medicine have special characteristics regarding the chemical structure and the mechanism of action, interfering directly or indirectly in the fungal cell, with fungistatic or fungicidal actions [34].

Among the existing antifungal drugs, the most widely used and known are polyenic, imidazolic, pyrimidine, sulfamide, benzofurenic and other compounds with varying degrees of success, such as iodides, thiosulfates, sulfides and tolnaftates. In the treatment of invasive fungal infections, classes of polyene antifungals (amphotericin B), azoles (fluconazole, voriconazole, ketoconazole, itraconazole, posaconazole), pyrimidines (5-fluorocytosine) and echinocandin, caspofungin, micafungin) are mainly used [35]. The increasing incidence of yeast infections, such as those present in the oral mucosa, has been a target of constant concern in the search for increasingly effective treatments and safer drugs. The use in the treatment and prophylaxis of antifungals such as fewer toxic azoles, especially fluconazole, has given rise to cases of resistance among susceptible yeast species.

The resistance of fungi to antifungal agents can be classified into clinical and microbiological resistance. The concept of clinical resistance is defined when there is a persistence or progression of a fungal infection even with the administration of the drug chosen as appropriate. In this case, "in vitro" tests may indicate the sensitivity of the agent to the antifungal. Usually, the occurrence of clinical resistance is associated with host, iatrogenic, pharmacological factors and factors related to the fungus virulence [36]. Microbiological resistance is a phenomenon in which the etiologic agent can develop in the presence of therapeutic concentrations of antifungals, a capacity verified "in vitro". Resistance can be intrinsic, primary or secondary, or extrinsic. This aspect is of real importance since we are increasingly faced with resistant yeasts, especially the "critical" species, highlighting *C. auris* and *C. haemulori*, whose findings should be immediately reported to the treatment team.

Intrinsic resistance is so called when no member of a species is sensitive to the antifungal, being primary, when in a species normally sensitive to an antifungal we find a resistant strain (without exposure to it) or secondary or acquired, when a previously sensitive strain develops resistance after exposure to a drug, due to phenotypic or genotypic changes [37]. The mechanism of resistance to antifungals by fungi, both for clinical or microbiological resistance, is involved with cellular, biochemical and/or molecular responses.

In the cellular mechanism, strains or sensitive specimens are exchanged for resistant endogenous ones, genetic alteration, a fact that guarantees secondary resistance, transient genetic expression and alteration in the cell type. Regarding the biochemical mechanism, phenotypic changes in fungi occur, allowing the absorption of the drug to be slower, altering the target site and increasing the excretion of the drug. The changes from the molecular point of view causing a genetic amplification to occur, mutations, among other modifications in the gene involved in the defense against the antifungal. In addition to these changes, another molecular alternative of resistance is the ability to form biofilms, an efficient physical barrier [36].

The greater phenotypic variability of *Candida* species, for example, together with the increased resistance of strains to antifungals, has assumed a prominent role as a clinical problem [9]. The different species of this yeast vary in sensitivity to antifungals on the market, a fact that shows the great importance of identifying and determining the minimum inhibitory concentration (MIC) [38].

Due to this aspect, the development of standardized methods of sensitivity "in vitro" is of vital importance and serve as guide to indicate the therapeutic choice, monitor the effectiveness of the antifungal and decrease the formation of resistant strains [39]. The appropriate choice of antifungal agent is, therefore, decisive in the therapeutic response of the animal. To this end, research that aims at determining the antifungal profile of the main yeasts isolated from dogs is of great therapeutic value [9].

The most used parameter for determining sensitivity to antifungals is the minimum inhibitory concentration (MIC), defined as the lowest concentration of an antifungal agent that inhibits the growth of the fungus [40]. From the MIC value, the yeast sample is classified according to the breakpoints established by international committees, which allows the fungus to be characterized as sensitive, intermediate, dose-dependent and resistant sensitivity [38]. For the detection of sensitivity/resistance to antifungals, highlight of use in therapeutic failures, we can count on several techniques, being "Gold standard" the method recommended by CLSI, the microdilution test.

The determination of the sensitivity of a fungus to antifungals can also be determined by commercial methods compatible with the tests recommended by CLSI [17, 37]. Sensitivity tests using a solid medium, such as the commercial method "E-test", are of real interest in several studies and in the laboratory routine. It is an excellent technique for determining the sensitivity to antifungals "in vitro", simple, easy to perform, with fast results, without the need for expensive or specialized equipment [41]. "E-test" is based on a combination of dilution and diffusion test concepts that directly quantify antifungal sensitivity. It consists of tapes containing pre-established concentrations of the antifungal agent, which are placed in a solid medium and with the yeast sample. When the tape is applied to the plate, immediate drug release occurs, thus the MIC is determined by the intersection of the inhibitory hyperbole formed by the growth of yeast in the plate (**Figure 18**).

Because the MIC values of "E-test" are directly proportional to the values referenced by the dilution CLSI, this method has a good correlation with this test. However, it may still present differences inherent to the process.

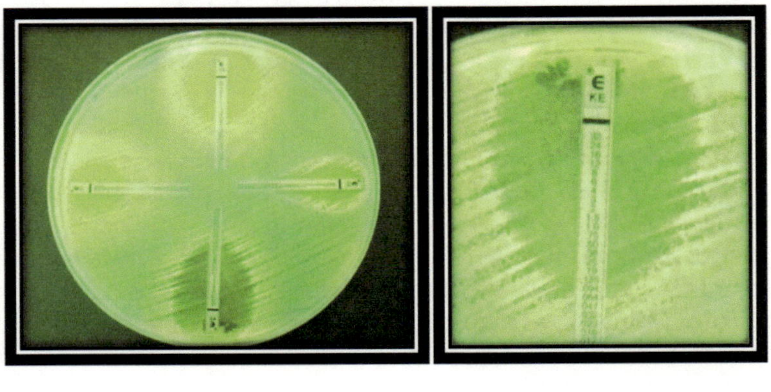

Figure 18.
"E-test" commercial method. MIC is determined by the intersection of the inhibitory area formed by yeast growth.

Antifungals	S (μg/mL)	SDD (μg/mL)	R (μg/mL)
Miconazol*	< 8	8–16	≥ 16
Cetoconazol	< 16	—	≥ 16
Fluconazol	≤ 8	16–32	≥ 64
Itraconazol	< 0,25	0.25–0.5	≥ 1
Voriconazol	≤ 1	2	≥ 4
Caspofungina	≤ 2	—	> 2

S: sensitive; SDD: dose dependent sensitivity; R: resistant; NS: not sensitive [38, 42].

Table 1.
Interpretation of the behavior of yeast strains against the concentration of antifungals (μg/ml).

The classification by this method determines the isolate as sensitive, dose-dependent and resistant (**Table 1**).

For tests to determine the antifungal profile, source control strains of the "American Type Culture Collection" (ATCC) are always used under identification, such as, for example, ATCC64548 (*C. albicans*) and ATCC777 (*C. dubliniensis*).

In the treatment of invasive fungal infections, classes of polyene antifungals (amphotericin B), azoles (fluconazole, voriconazole, ketoconazole, itraconazole, posaconazole), pyrimidines (5-fluorocytosine) and echinocandin, (caspofungin, caspofungin, micafungin) are mainly used [35].

For the systemic treatment of yeasts, we can use Amphotericin B, in the most varied forms (liposomal, suspension of lipid complexes). Nystatin can be used orally or in suspension, ointments and creams (as for example, in cases of oral candidiasis). In animals, the use of each of these antifungals is quite varied and their recommendation and dose will depend a lot on the etiological agent in question and the side effects that can be generated.

When analyzing the profile of sensitivity to antifungals compared to isolates from the oral cavity of dogs (mucosa that has greater transmissibility to humans), the best active antifungals found in the veterinary are ketoconazole and voriconazole. Ketoconazole is still widely used in clinics and pet shops, mainly, topically. For the treatment of candidiasis in small animal clinics, ketoconazole is one of the most frequently used drugs, as it has a broad spectrum of activity, encompassing several species of *Candida* spp. and dermatophytes. From isolates from the oral cavity of dogs it shows high sensitivity between yeasts and has several presentations for veterinary use, representing an economically viable alternative, however, due to its toxicity, the trend is disuse [9, 17].

Voriconazole has a broad spectrum of activity and a potent "in vitro" action. Its mechanism of action is like other azole antifungals, inhibiting the enzyme 14 alpha-demethylase, dependent on cytochrome P-450, essential for the ergosterol biosynthesis. It can be indicated as a good alternative to replace ketoconazole, however its cost is high. This drug is also used for the treatment of systemic mycoses, mainly in candidiasis, aspergillosis and cryptococcosis in debilitated, immunosuppressed patients or in cases of resistance to another antifungal [43]. The antifungals fluconazole, itraconazole and miconazole are also routinely applied in the veterinary clinic, used indiscriminately in the treatment of mycosis suggestively diagnosed. However, resistance to these drugs has increased, so their use should be more cautious.

Candida zeylanoides, for example, is a relatively rare yeast in humans and animals. In humans it has been reported from skin, nails and blood isolation, considered an opportunistic pathogen, also involved in cases of endocarditis in an

HIV-positive patient [44]. Samples of this yeast were isolated for the first time in the oral cavity of stray dogs and demonstrated significant resistance to the antifungal fluconazole. In addition to this species, *Candida krusei* also obtained partial results of resistance to this antifungal, as well as yeasts of the genus *Trichosporon* spp. [9].

Itraconazole is a synthetic triazole derivative with a wide spectrum of action, widely used in the treatment of superficial mycoses by candidiasis, malasseziosis and in systemic mycoses. When used orally right after a meal, its bioavailability is maximum, with biphasic elimination. This antifungal has also been used successfully in dogs with mycotic rhinitis and in systemic mycoses, such as blastomycosis. However, its use in dogs can lead to skin rashes and, in high dosages, it can cause anorexia and increased plasma concentration of alkaline phosphatase and aminotransferase enzymes [43].

In addition, species isolated from the oral cavity of dogs (especially *Candida albicans* and *C. tropicalis*) have shown dose-dependent sensitivity to itraconazole. Yeasts of the genus *Trichosporon* also isolated from this active site, show medium resistance to fluconazole and significant resistance to itraconazole, which reveals concern about the use of these drugs in the treatment of candidiasis and triconosporoses in dogs [9, 19].

In the veterinary medical clinic, miconazole is commonly indicated for the treatment of dermatophytosis, malasseziosis and candidiasis. However, yeasts of the genus *Trichosporon* and *Malassezia pachydermatis* isolated from the oral cavity of dogs show important resistance to this antifungal. Different for *Candida* yeasts, in which the antifungal profile demonstrates sensitivity to miconazole [9, 19].

Caspofungin is an antifungal with an inhibitory action on the cell wall of the echinocandin group, important in human medicine as an alternative for the treatment of isolates resistant to fluconazole [45]. Against yeasts isolated from the oral cavity of dogs, yeasts of the genus *Trichosporon* and of the genus *Malassezia* demonstrate significant resistance to this antifungal, resistance also demonstrated to a lesser extent by the species *Candida parapsilosis* [9, 19].

In cases of systemic infections, affecting different species of animals, the use of amphotericin B, a drug that acts on the fungal cell membrane, has efficiency against strains of *Candida* spp. However, due to the high cost and serious side effects, such as hepatotoxicity, nephrotoxicity, myelotoxicity and cardiotoxicity, this medication is seldom used [17].

Due to the great similarity between the fungal cell and the host cell, the action of antifungals presents relatively high toxicity. Thus, there is a need for research for the best choice of antifungal, based on the most appropriate therapeutic response and on the sensitivity profile of yeast against antifungal floodgates, seeking as well to minimize the side effects that can be generated with the use of more drugs needed in cases of therapeutic failure [43].

When information is obtained that a street animal, which in general is a dog that, has never received therapeutic treatment based on antifungal, presents positive isolation for resistant yeasts, it is assumed that environmental yeasts are undergoing an important primary resistance or that the ecological niche in which that animal lives is contaminated by resistant microorganisms originating from direct or indirect human contamination.

Corroborating this fact, we must consider the excessive use of pesticides in the environment and mycoherbicides (placed in plantations, vegetable gardens, and in the soil itself), have a chemical constitution like azoles, thus representing a strong selective pressure for the emergence of strains resistant. This question of possible environmental contamination and fungal resistance is already discussed for other yeast species, such as *Cryptococcus* spp. and medical mycology becomes an important issue.

The growing data on increased resistance of fungi against antifungal drugs have been causing great concern for human and veterinary doctors. Although data on resistance to antifungals from yeasts isolated from dogs are scarce, their importance is notorious, directly associated with the therapeutic success of these animals particularly important for society and human health (physical and mental).

Therefore, the ideal therapeutic choice, for both humans and animals, should be based on prior identification of the agent and, if possible, the use of techniques for determining the sensitivity profile of the etiologic agent against antifungals.

9. Considerations

The oral cavity is an extremely important anatomical area of dogs, considered as one of the determining factors in the longevity of this animal's life. To reduce therapeutic failures and guarantee the perfect health condition of this system, knowledge of the existing microbiota is essential, but it is still scarce.

We can then ask ourselves: Why are recurrent fungal infections more and more frequent in dogs? What is the relationship between the microbiota of dogs and their respective owners? And what is the relationship of resistance to fungal infections between these species?

Possible answer to these questions could be founded in this chapter, as well as the beginning of the knowledge of the main yeasts found in the oral cavity of dogs, their clinical importance and profile of resistance to the main antifungals used in the practical routine of veterinary medicine.

Acknowledgements

The authors would like to thank Professor Renata Mangione for the translation into English and the Brazilian Official Institutes (FAPESP, CNPQ and CAPES).

Author details

Claudete Rodrigues Paula[1*], Bianca Silva Navarro[2], Mário Mendes Bonci[1] and Diana Costa Nascimento[3]

1 School of Dentistry, University of Sao Paulo (USP), Sao Paulo, Brazil

2 Clinical Veterinarian Animal Facility, Zoetis Industria de Produtos Veterinários LTDA, Campinas, Brazil

3 Pontifical Catholic University of Campinas (PUC), Campinas, Brazil

*Address all correspondence to: crpmicol@uol.com.br

References

[1] KURTZMAN, Cletus; FELL, Jack W.; BOEKHOUT, Teun (Ed.). **The yeasts: a taxonomic study.** Elsevier, 2011.

[2] GOMPERTZ, Olga Fischman; GAMBALE, Walderez; CORRÊA, Benedito; PAULA, Claudete Rodrigues. Estrutura, morfologia, reprodução e taxonomia dos fungos. In: TRABULSI, Luiz Rachid; ALTERTHUM, Flavio (Ed.). **Microbiologia.** São Paulo: Atheneu.2015.

[3] WATKINSON, Sarah C.; BODDY, Lynne; MONEY, Nicholas. **The fungi.** Academic Press, 2015.

[4] LOPES, Katia Regina Freire. Considerações sobre a importância do cão doméstico dentro da sociedade humana. **Acta Veterinaria Brasilica**, v. 6, n. 3, p. 177-185, 2012.

[5] NAGASAWA, Miho; MOGI, Kazutaka; KIKUSUI, Takefumi. Attachment between humans and dogs. **Japanese Psychological Research**, v. 51, n. 3, p. 209-221, 2009.

[6] DENG, Ping; SWANSON, Kelly S. Gut microbiota of humans, dogs and cats: current knowledge and future opportunities and challenges. **British Journal of Nutrition**, v. 113, n. S1, p. S6-S17, 2015.

[7] ROSE, Anthony H.; HARRISON, J. Stuart. **The yeasts. V. 1. Biology of yeasts. V. 2. Physiology and biochemistry of yeasts. V. 3. Yeast technology**. Academic Press, 1971.

[8] PAULA, Claudete Rodrigues; RUIZ, Luciana da Silva. Importância das leveduras em infecções nosocomiais: marcadores fenotípicos e genotípicos. In: ZAITZ, Clarisse et al. **Compêndio de Micologia Médica**. Rio de Janeiro: Guanabara Koogan. 2010.

[9] NAVARRO, B. S. et al. Antifungal sensitivity and species of yeasts in oral mucosa of street mixed-breed dogs. **Journal de Mycologie Médicale**, v. 30, n. 4, p. 101010, 2020.

[10] FLAUSINO, Gilberto et al. Isolation and characterization of *Cyniclomyces guttulatus* (Robin) Van Der Walt and Scott, 1971 in dogs in Brazil. **Current microbiology**, v. 65, n. 5, p. 542-546, 2012.

[11] SEYEDMOUSAVI, Seyedmojtaba et al. (Ed.). **Emerging and epizootic fungal infections in animals**. Springer, 2018.

[12] FIGUEREDO, Luciana A. et al. Biofilm formation of *Malassezia pachydermatis* from dogs. **Veterinary microbiology**, v. 160, n. 1-2, p. 126-131, 2012.

[13] DA SILVA MACHADO, Mauro Luís; APPEL, Carin Elisabete; FERREIRO, Laerte. Dermatófitos e leveduras isolados da pele de cães com dermatopatias diversas. **Acta Scientiae Veterinariae**, v. 32, n. 3, p. 225-232, 2004.

[14] GASPARETTO, André et al. Produção de biofilme por leveduras isoladas de cavidade bucal de usuários de prótese dentária. **Acta Scientiarum. Health Sciences**, v. 27, n. 1, p. 37-40, 2005.

[15] LACAZ CS, PORTO E, MARTINS JEC, HEINS-VACCARI EM, MELLO NT. Tratado de micologia médica Lacaz. São Paulo: **Savier**; 2002.

[16] GOMPERTZ Olga Fischman, et al. Micoses oportunísticas e outras micoses. In: Trabulsi LR, Alterthum F. **Microbiologia.** 4. ed., São Paulo: Atheneu; 2008b. Capítulo 70. p. 525-30.

[17] BRITO, Erika Helena Salles de et al. Candidose na medicina veterinária: um enfoque micológico, clínico e

terapêutico. **Cienc. Rural**, Santa Maria, v. 39, n. 9, p. 2655-2664, Dec. 2009.

[18] KONEMAN, Elmer et al. Diagnóstico microbiológico: texto e atlas colorido. In: **Diagnóstico microbiológico: texto e atlas colorido**. 5. ed. Rio de Janeiro: Médica e Científica; 2001.

[19] NAVARRO, Bianca Silva. Fenótipos e perfis de sensibilidade aos antifúngicos de leveduras isoladas da mucosa oral de cães da cidade de Campinas, São Paulo. 2016. Dissertação (Mestrado em Diagnóstico Bucal) - Faculdade de Odontologia, University of São Paulo, São Paulo, 2016.

[20] REIS, Emily Correna Carlo; BORGES, Andréa Pacheco Batista; DEL CARLO, Ricardo Junqueira. Regeneração periodontal em cães. **Ciência Rural**, v. 41, n. 12, p. 2128-2136, 2011.

[21] VENTURINI, Michele Alice Françoise Anita. Estudo retrospectivo de 3055 animais atendidos no ODONTOVET® (Centro Odontológico Veterinário) durante 44 meses. 2007. Dissertação (Mestrado em Clínica Cirúrgica Veterinária) - Faculdade de Medicina Veterinária e Zootecnia, University of São Paulo, São Paulo, 2007.

[22] KYLLAR, M.; WITTER, K. Prevalence of dental disorders in pet dogs. **VETERINARNI MEDICINA-PRAHA-**, v. 50, n. 11, p. 496, 2005.

[23] DUBOC, Marcela Vieira. Percepção de proprietários de cães e gatos sobre a higiene oral de seu animal. 2009. 62 f. Dissertação (Mestrado em Ciências Veterinárias) - Instituto de Veterinária, Universidade Federal Rural do Rio de Janeiro, Seropédica - RJ, 2009.

[24] PAIVA, Alessandra Costa et al. Eficácia dos coadjuvantes de higiene bucal utilizados na alimentação de cães. **Arq. Bras. Med. Vet. Zootec.**, Belo Horizonte, v. 59, n. 5, p. 1177-1183, Oct. 2007.

[25] GARCIA Claudia Zeferino, et al. Doença periodontal em cães. **Rev Cient Eletr Med Vet.** 2008.

[26] Brito Erika H.S. et al. The anatomical distribution and antimicrobial susceptibility of yeast species isolated from healthy dogs. **Vet J** 182(2). 2008.

[27] JIN Yipeng, LIN Degui. Fungal urinary tract infections in the dog and cat: a retrospective study.**Journalof the American Animal Hospital Association**. 2005

[28] PINTO, Priscila Natália, et al. Atlas de Micologia Médica Veterinária. **Cad. técn. Vet. Zoot.** *;* (94): 109 p, dez. 2019.

[29] BENTUBO, Henri Donnarumma Levy; GAMBALE, Walderez; FISCHMAN, Olga. Caracterização laboratorial e comportamento cromogênico de leveduras do gênero Trichosporon. **Revista Brasileira de Pesquisa em Saúde/Brazilian Journal of Health Research**, 2013.

[30] FISHER, Fran; COOK, Norma B.; FIORAVANTI, Irma. Micologia: fundamentos e diagnóstico. In: **Micologia: fundamentos e diagnostico**. 2001. p. 337-337.

[31] GANDRA, Rinaldo Ferreira, et al. Malassezia spp. in Acoustic Meatus of Bats (*Molossus molossus*) of the Amazon Region, Brazil. **Mycopathologia**, 165. 2008.

[32] HEITMAN, Joseph et al. **Cryptococcus: from human pathogen to model yeast**. USA: ASM press. 2010.

[33] PAULA, Claudete Rodrigues; RUIZ, Luciana da Silva.Técnicas para o diagnóstico precoce das infecções por leveduras do gênero *Candida*. In: ZAITZ, Clarisse et al. **Compêndio de**

Micologia Médica. Rio de Janeiro: Guanabara Koogan. 2010.

[34] APPELT, Carin Elisabete; CAVALCANTE, Liziane Ferraresi Holanda. Malassezia pachydermatis em cães e sua susceptibilidade aos antifúngicos azóis: revisão de literatura. **Vet Foco**. 6(1): 21-28, jul.-dez. 2008.

[35] PAPPAS, Peter G. et al. Clinical practice guidelines for the management of candidiasis: 2009 update by the Infectious Diseases Society of America. **Clinical infectious diseases**, p. 503-535, 2009.

[36] CURY AE. Resistência a antifúngicos utilizados em micologia médica. In: Zaitz C, Campbell I, Marques AA, Ruiz LRB, Framil VMS. Compêndio de micologia médica. 2. ed. Rio de Janeiro: **Guanabara Koogan;** 2010.

[37] MELHEM Marcia de Souza Carvalho, SZESZS Maria Walderez. Testes de suscetibilidade a drogas antifúngicas. In: Zaitz C, Campbell I, Marques AA, Ruiz LRB, Framil VMS. **Compêndio de micologia médica.** 2ª ed. Rio de Janeiro: Guanabara Koogan; 2010. p. 406-422.

[38] CLSI - Clinical and Laboratory Standards Institute. Reference method for broth dilution antifungal susceptibility testing of yeasts. Approved Standard M27-A3. Wayne, PA: CLSI; 2008.

[39] NUNES, Emmanuel Borges et al. Perfil de sensibilidade do gênero Candida a antifúngicos em um hospital de referência da Região Norte do Brasil. **Rev Pan-Amaz Saude**, Ananindeua, v. 2, n. 4, p. 23-30, dez. 2011.

[40] EUCAST - European Committee on Antibiotic Susceptibility. Definitive Document E Def 7.1: method for the determination of broth dilution MICs of antifungal agents for fermentative yeasts. **Clin Microbiol Infect**. 2008.

[41] MATSUMOTO Flávia E. et al. Antifungal susceptibility of bloodstream yeasts isolated at a public children's hospital in Brazil: comparison of the "Etest®" and the AFST–EUCAST microdilution method. **Can J Microbiol**53(12). 2007.

[42] DALAZEN, Daniela et al. Comparação do perfil de suscetibilidade entre isolados clínicos de Candida spp. orais e vulvovaginais no Sul do Brasil. **J. Bras. Patol. Med. Lab.**, Rio de Janeiro, v. 47, n. 1, p. 33-38, Feb. 2011.

[43] DE OLIVEIRA, Márcia Nobre et al. Drogas antifúngicas para pequenos e grandes animais. **Ciência Rural**, v. 32, n. 1, p. 175-184, 2002.

[44] KHOSRAVI et al. Evaluation of the pathogenicity of *Candida zeylanoides* in BALB/c mice. **Turk J Vet Anim Sci,** 2013.

[45] NAGAPPAN Vijayalakshmi, BOIKOV Dina, VAZQUEZ Júlio A. Assessment of the "in vitro" kinetic activity of caspofungin against *Candida glabrata*. **Antimicrob Agents Chemother**54 (1). 2010.

Section 3

Oncology

Chapter 6

Small Animals Gut Microbiome and Its Relationship with Cancer

Tatiane Moreno Ferrarias Epiphanio and Andreia A.F. Santos

Abstract

This chapter aims to discuss recent developments in understanding the small animal gut microbiome's relationship with cancer, focusing on animals as well as a model for studying humans. Based on multidirectional interactions between the microbiome, the environment and the epigenetically/genetically vulnerable host, it intends to address the mechanisms by which microorganisms can contribute to carcinogenesis describing the roles of the microbiome directly in the pathogenesis of the disease through complex interactions between the microbiome and the host's metabolic and immune systems. The feasibility for developing new cancer diagnostic and prognostic methodologies plus treatments based on small animals' microbiome profiles are reviewed.

Keywords: gut microbiome, carcinogenesis, therapy, dog, cancer

1. Introduction

Much recent medical research focuses on understanding the influences of the microbiome on host health and disease progression such as in inflammatory, metabolic, autoimmune and oncologic diseases [1].

In order to introduce the reader to this chapter, it is essential to clarify some common terms such as microbiota, metataxonomics, microbiome and metagenome. The microbiota is defined as the assemblage of living microorganisms present in a certain environment and is composed by bacteria, archaea, fungi, algae and small protists [2, 3]. Metataxonomics defines the high-throughput process used to taxonomically identify microorganisms in the environment and characterize the entire microbiota, creating a metataxonomic tree [2]. The definition of microbiome includes not only the microorganisms community, but also their "theatre of activity" that involves the whole spectrum of molecules produced by them, including their structural elements (nucleic acids, proteins, lipids, polysaccharides), metabolites (signaling molecules, toxins, organic, and inorganic molecules), and molecules produced by coexisting hosts and structured by the environmental conditions [3]. It stands out that all mobile genetic elements, such as phages, viruses, and extracellular DNA should be included in the term microbiome but are not a part of microbiota [3]. Lastly, the term metagenome refers only to the collection of genes and genomes of members of a microbiota [2].

In humans, as well as in small animals, these complex communities of microbes inhabit predominantly the gastrointestinal tract and oral cavity, but other exposed tissues, such as skin, breast, respiratory and urinary tract, can also harbor unique bacterial communities [4–9]. The host microbiota and immune system must communicate to maintain a balance between tolerance and activation, otherwise a dysbiotic state can be established and may incite or sustain diseases, such as cancer [10]. Epidemiological associations of abnormal microbiome with gastric, esophageal, hepatobiliary, pancreatic, lung, colorectal, lymphoma and other human and canine cancers have been previously established [11–13].

Neoplastic processes are the leading cause of death in adult dogs [14]. The annual cancer incidence rate is 381 per 100,000 dogs with 4 million new cancer cases per year, similar to the reported rate in humans (454 per 100.000) with 18 million new cancer cases annually [15–17]. In these species, naturally occurring cancers share many features, including clinical presentation, biological behavior, histological features, tumor genetics, and treatment response [18, 19]. Coelho and colleagues showed that the dog gut microbiome has a higher taxonomic and functional overlap with the human gut microbiome than pigs or mice and concluded that findings in dogs may be predictive of human microbiome results [20]. In addition, companion animals represent a special human experimental model in microbiomic investigations due to the exchange of microbes between humans and their pets [21].

In humans, there are some reviews involving microbiome and cancer, but they are scarce in veterinary medicine, with the most reviews covering the microbiome gastrointestinal tract and other diseases [22, 23]. The present article reviews the current status of comparative oncology approaches in human and small animals in the field of microbiome with special focus on carcinogenesis, relationship between specific microbiomes as well as the feasibility of new cancer diagnostic tools and therapies based on microbiome profiles.

2. Human and small animal microbiomes

The host's first major exposure to a complex microbiota occurs during birth through contact with the maternal microbiome, which represent a primary mechanism for the intergenerational microbiota transfer in mammals and, afterwards, bacterial colonization progresses from childhood to adulthood [24]. The microbiota development is limited to its niches by the host's immune system, along with the host's chronological development, providing early modulation of the host's physiological development and functions of nutrition, immunity and resistance to pathogens at all ages [24].

The most important group of organisms in microbiome studies is called the dynamic symbionts, whose symbiotic nature may vary along a spectrum from mutualism and commensalism to parasitism and amensalism [25]. Usually, microbes perform synthetic or catabolic metabolic activity through direct microbe-host interactions. Catabolism and bioconversion of compounds from the diet make nutrients more available to the host through the processes of fermentation, hydrolysis, metabolism of drugs and toxins, among others. Some microbiota members can synthesize important cofactors or bioactive signaling molecules such as vitamins and active amines. In addition, this can trigger changes in the host's gastrointestinal epithelial and immune responses [26].

The combination of factors such as age, genetics, physiological status (including innate and adaptive immune system), lifestyle, diet, host environment and disease status can result in variation in microbiomes between hosts [27]. Human gut microbiota is extremely diverse, with an estimated 1,000 bacterial species in the gut with 2,000 genes per species yields an estimate of 2,000,000 genes, which is 100 times the commonly estimated 20,000 human genes [27]. In dogs, gut microbiome contains around 1,200,000 genes [20] and recent studies suggest that canine and feline gut fecal microbial phylogeny (e.g. predominance of Firmicutes, Bacteroidetes, Proteobacteria and Actinobacteria) and functional capacity (e.*G. major* functional groups related to carbohydrate, protein, DNA and vitamin metabolism, virulence factors and cell wall and capsule) are similar to those of the human gut [28].

3. Microbiome and carcinogenesis

Cancer is a complex disease, in which cumulative genetic, epigenetic physiological, immunological and biochemical changes occur incessantly in the tumor tissue, contributing to the complexity of the understanding, treatment and management of the disease. It is estimated that microorganisms could be associated with 15–20% of cancers [29].

As mentioned, the microbiota has an essential role in host health, in which a beneficial relationship is established, however, dysbiotic states can trigger several diseases, including cancer. Scott and colleagues proposed that in the etiopathogenesis of cancer, dysbiosis should be considered a persistent exit of the host microbiome from the health-associated homeostatic state (consisting of mutualists and commensals), towards a cancer promoting and/or sustaining phenotype (parasitism or amensalism) [25]. Currently, metataxonomic and metagenomics studies have documented and compared the diversity and abundance of microbes in different parts of the body between healthy and diseased patients. In veterinary medicine, it has been demonstrated a significant difference in the microbial communities in dogs with intestinal and multicentric lymphoma and with colorectal tumors comparing to healthy dogs [12, 13, 30]. However, these studies cannot distinguish whether some alterations in microbiota are causes or effects of cancer, describing only the different microbial communities found among the study groups.

The microbiome causative role has been demonstrated by controlled pre-clinical studies utilizing germfree (i.e., devoid of any microbiota) mouse models colonized with selected bacteria. For example, several family members of Enterobacteriaceae, including *Escherichia coli*, harbor an island of polyketide synthase (pks) pathogenicity that synthesizes a genotoxin called colibactin [31]. In an experimental study, knockout mice for IL-10 were mono-associated with two strains of *E. coli* that were pks + or Δpks (with and without pks, respectively) and treated with pro-carcinogenic azoxymethane to induce colorectal tumors to demonstrate that pks play a causal role in tumorigenesis [31]. All mono-associated pks + mice developed invasive carcinoma, in contrast, none of the Δpks mono-associated mice exhibited full invasion [31]. This result suggests that the presence of *E. coli* pks accelerates the progression from dysplasia to invasive carcinoma through the genotoxicity of colibactin, an example of pathway of the microbiota-associated carcinogenesis process.

In a recent consensus on the human microbiome role in carcinogenesis, expert opinion was that the microbiome is one apex of a tripartite, multidirectional interactome alongside environmental factors (such as diet, obesity) and an epigenetically/genetically vulnerable host that combine to cause cancer [25]. Gastrointestinal microbiome, which comprises 99% of the microbial mass, not only has the greatest both local and long-distance effects on overall health and metabolic status, but it is also the best investigated microbiome and serves as a model for understanding host–microbiota interactions and disease [32]. Due to its location, gut microbiome has been well studied as a contributor to colorectal carcinogenesis [33]. Other organs with a well-characterized microbiome include the skin and the vagina [34, 35]. The microbiome of each organ is distinct suggesting that effects on inflammation and carcinogenesis are likely to be organ specific. Although many organs (e.g. liver and brain), does not have a known microbiome, they may be exposed to pathogen-associated molecular patterns (PAMPs) and bacterial metabolites through anatomical links with the gut [32, 36].

For a better understanding of the microbiome role in carcinogenesis it is important to recognize that bacteria can be found in the tumor tissue itself, in normal adjacent tissue and in tumor sites, such as intestine and genitourinary tract, with overlap between these sites (**Figure 1**). According to Picardo, the microorganisms inside, adjacent and distant from the tumor can play a role in cancer development and progression and interactions between these microbial populations together with the indirect gut microbiome effects have the potential to influence the disease development [37].

At the molecular level, the mechanisms by which microorganisms can contribute to carcinogenesis are multiple and varied, which may broadly be categorized into genomic integration and genotoxicity (by a direct oncogenic effect of microorganisms and their products); promotion of immunological modifications (which disrupts host cancer immunosurveillance through the induction of pro-inflammatory and immunosuppressive pathways); and metabolic reprogramming (by altering circulating metabolites which become pro-carcinogenic and by stimulating the synthesis of trophic factors for cancer cells by the host). Many of these actions can harm the host indirectly, as microbes optimizes conditions for their survival may result in a final common pathway of prolonged host cell survival, enhanced replicative capacity and dedifferentiation [25, 33]. These mechanisms converge to hallmarks of cancer [38] and will be described in more detail below.

3.1 Genomic integration

Although the microbiome viral communities have not been studied as much as the bacterial community, the virus's ability to integrate into the host genome is a causal mechanism of cancer both in dogs and humans.

A remarkable example is the human papilloma virus (high-risk HPV 16 and 18) and its association with human cervix cancer. The key event of HPV-induced carcinogenesis is the integration of two HPV genes (E6 e E7) into the host genomic DNA [39]. In proliferating cells of the basal layer of the uterine cervix, the viral

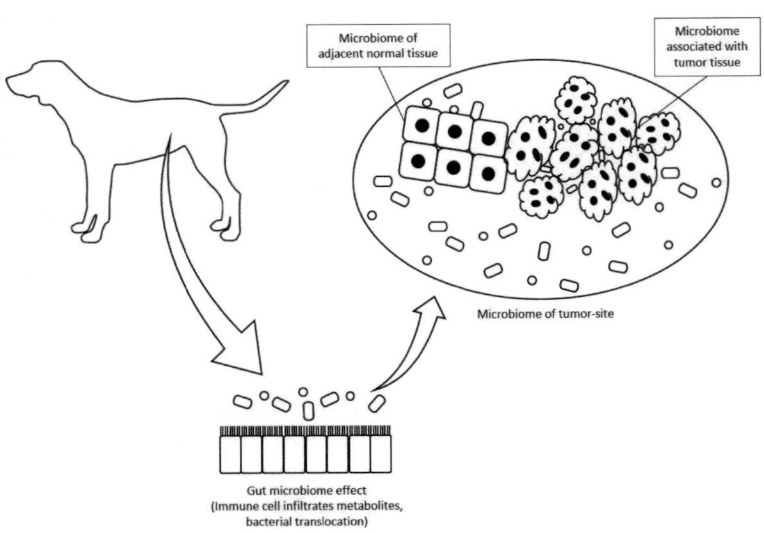

Figure 1.
The relationship between tumor and microbiomes (adapted from Picardo et al., 2019) [37].

genome persists as episomes, replicates in the suprabasal cells and can infiltrate deeper layers [40]. The HPV E6 and E7 genes are regularly present and expressed in the tumor tissue [40]. Their expression and the loss of expression of the E2 region (which negatively regulates E6 and E7) in the integrated HPV genomes cause the disruption of tumor suppressor genes that result in dysregulation of cell growth and inhibition of apoptosis [41]. Therefore, the overexpression of these viral genes synergistically acts to immortalize host cells, a cancer hallmark.

In dogs, investigations with canine papillomavirus (CPVs) have been limited to the association of different CPV genotypes with neoplastic lesions. Up to now, 20 CPVs types have been reported [42]. In skin, most genotypes of CPVs cause benign lesions, such as warts and pigmented/viral plaques or papillomas, which are self-limiting lesions such as those of oral papillomatosis [43].

Dogs that develop extensive papillomatosis may also be predisposed to oral squamous cell carcinoma (SCC) [44]. The detection of CPVs in malignant epithelial lesions is increasing in recent years [42, 45, 46]. CPV 1, 2, 3, 7, 12, 16, and 17 have been reported to cause epithelium neoplastic transformation. In a retrospective study, Thaiwong and colleagues (2018) described 7 dogs bearing benign papillomas associated with CPV1 and also the histological evidence of CPV1 causing malignant transformation of carcinoma *in situ* (ISC) and SCC. Later, the same group showed the expression of p53 and p16 proteins in cells infected with CPV1 in benign papillomas and lesions that progressed to SCC [42].

In a recent retrospective study, CPVs were successfully detected in 11 skin tissue samples and 4 oral tissues obtained from a cohort of canine papillomas and SCCs by PCR and through the detection of intralesional viral antigens using immunohistochemistry [46]. After sequencing, CPV 1, 2 and 6 were detected in the benign lesions, while CPV 9, 15 and 16 were detected in the SCCs, highlighting the risk of these genotypes in the induction of epithelial carcinogenesis [46].

The first report of chromosomal integration of CPV 16 into the host genome was detected in a sample of squamous cell carcinoma, raising the possibility that CPV 16 may be a potential type of high-risk canine papillomavirus [47]. However, the CPVs oncopathogenesis should be further investigated.

3.2 Genotoxicity

The gut microbiota is mainly composed of bacteria, many of which contain toxin-producing strains that can have carcinogenic effects through interfering with the cell cycle regulation, cell growth or directly damaging the host's DNA [48]. Pathogenic bacteria strains produce protein toxins to meet their survival needs, but these bacterial defense factors perturb the host equilibrium and affect tumor suppressor genes or oncogenes and promote host genome instability [49].

Among the large number of bacterial protein toxins, two genotoxins are well known for directly affecting the host's DNA integrity in the host organism target cells: cytolethal distending toxin (CDT), which is produced by several gram-negative pathogenic bacteria (e.g. *E. coli, Shigella dysenteriae, Campylobacter jejuni, Helicobacter sp.*) [50], and colibactin toxin (produced by *E. coli* strains); both trigger double-strand DNA breaks in host cells contributing to carcinogenesis [50, 51]. CDT exerts a pro-carcinogenic effect mainly because it presents a DNase activity. After binding to the host cell membrane, CDT suffers receptor-mediated endocytosis, proceeds to the endoplasmic reticulum and is translocated to the nucleus, where promotes cytotoxicity [50]. Cell CDT intoxication induces DNA damage, which results in the stopping of target cells in the G1 and/or G2 phases of the cycle and activation of DNA repair mechanisms [50]. Subsequently, normal cells that fail to repair the damage and survive the acute phase of CDT intoxication acquire the cancer hallmark of cellular senescence or undergo apoptosis via the DNA host damage checkpoint pathways [52]. This chromosomal instability supports the notion that CDT might promote tumor initiation and progression [53].

Colibactin-producing *E. coli* colonize frequently the colon mucosa of patients with human colorectal cancer (CRC) being implicated in carcinogenesis and tumor progression [54, 55]. This genotoxin is found in 55–67% of human colorectal cancer compared to less than 20% of controls [31, 54]. Understanding of colibactin's chemical structure and biological activity is limited, but recent studies have shown that these toxins are powerful DNA-damaging agents acting via alkylation and DNA cross-linking, whose lesions activate the DNA damage checkpoint pathway and cells present signs of incomplete DNA repair, G2/M cell cycle arrest and chromosomal instability [51, 56]. In addition, colibactin also supports tumor growth by inducing a secretory phenotype associated with senescence through growth factors secretion [57].

In veterinary medicine, Feng and colleagues identified *E. coli* strains encoding colibactin cytotoxic necrotizing factor (CNF) in the rectal swabs and extra-intestinal samples of macaques, whose can cause clinical and subclinical diseases [58]. Genotoxins in companion animals have not been identified so far, but the fecal microbiota composition in dogs with colorectal epithelial tumors was different from that of control dogs, where Enterobacteriaceae, Bacteroides, Helicobacter, Porphyromonas, Streptococcus and Fusobacteriaceae were overrepresented in those with tumors [13]. Thus, studies are still needed to identify genotoxins produced by bacteria to help understand the carcinogenesis of canine colorectal epithelial tumors.

3.3 Immunological modifications

There is a well-defined bidirectional interaction between the immune system and gut microbiome, playing a role in the entire organism physiology [59]. The gut microbiota is essential for normal development of innate and adaptive immunity at several levels (demonstrated by studies using germ-free mice) and the immune system regulates colonization and abundance of microbiome species, as well as the response to commensal bacteria [60–63].

The host microbiota and the immune system must communicate to maintain a balance between the inflammatory response activation and the immune tolerance preservation [64]. For this, the gut bacterial population presents both a protective and harmful interface. Unlike opportunistic bacteria, other commensals, such as *Bifidobacterium infantiles* and *Faecalibacterium prausnitzii*, induce the development of regulatory T cells that prevent an inadequate immune response and protect the host against intestinal pathogens [65]. A lack of control in pro-inflammatory and anti-inflammatory bacteria (causing an imbalance between Th17 and T-regulatory cells) establishes a dysbiotic state [65, 66]. Immunological intolerance results in a loss of homeostasis that can promote a pro-neoplastic inflammatory environment through chronic inflammation, immune evasion and immune suppression [32, 67].

The gut mucosa consists of a single epithelial cell layer with intraepithelial lymphocytes that facilitates the interaction of bacterial with immune system. The epithelial line contains Paneth cells that secrete anti-microbial molecules and goblet cells that secrete mucus to lubricate the intestinal contents and protect the epithelium, while on the skin, keratinocytes regulate the microbes by secreting antibacterial peptides [68]. The lamina propria is below the mucous layer, which contains a series of other immune cells (including antigen presenting cells and CD4 + and CD8 + T and B cells). This lymphoid tissue is the most important component of body's immune system, capable of influencing immune responses both locally and systematically [1]. Microbe is detected using pattern recognition receptors (PRRs) represented by Toll-like receptors (TLRs) and NOD-like receptors (NLR) [69].

These are widely expressed in intestinal epithelial cells, as well as in intestinal macrophages and dendritic cells. PRRs can either control the microbiota through antibacterial mediators and thus suppress cancer, or they can promote resistance to cell death - a hallmark of cancer [38].

The systemic immune system is prepared (at the epigenetic or transcriptional level) to enact a robust response in the presence of pathogenic bacteria leading to proinflammatory immune responses or to maintain a non-inflammatory state in the absence of threat [70]. A state of disruption of the delicate balance of commensal bacteria (dysbiosis), which is characterized by a less stable microbiota, increases the potential of opportunistic pathogenic bacteria growth [71]. As seen, dysbiosis can promote impaired local, loco-regional and systemic immune responses, being able to generate a profound inflammatory state, both locally and systemically. This process is outlined in **Figure 2**.

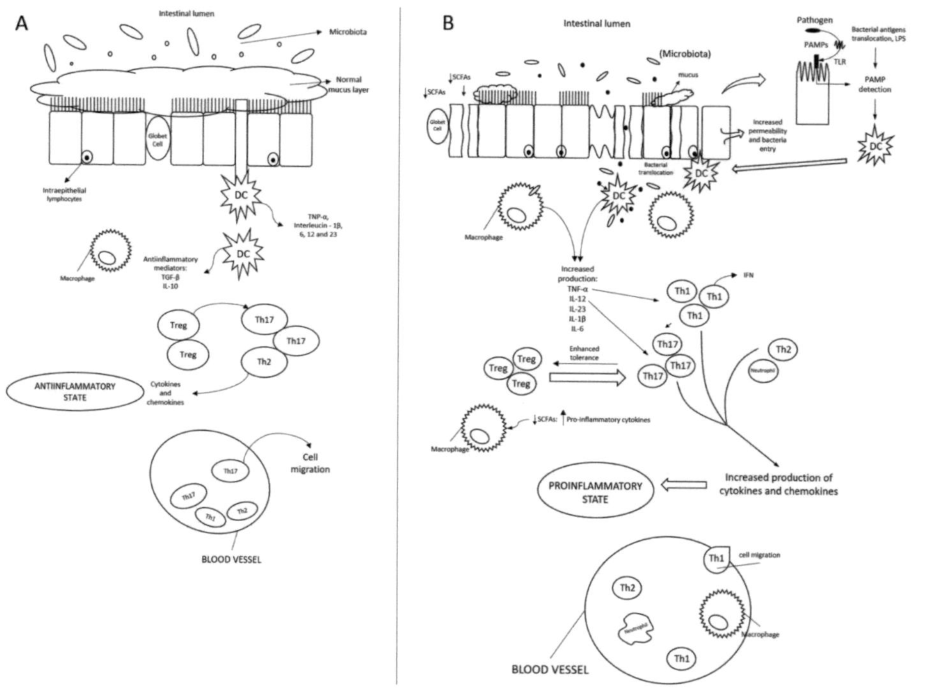

Figure 2.
The gut immune system in healthy and dysbiotic microbiome. (A): In healthy dogs, the lamina propria normally contains immune cells and secreted cytokines. These include anti-inflammatory mediators (transforming growth factor β [TGF-β] and interleukin (IL) -10) that down-regulate immune responses, limit excessive entry of intestinal microbiota and defend against pathogens; and noninflammatory defenses such as phagocytosis by macrophages, that assist in defending against bacteria entering the lamina propria. A homeostatic balance is maintained between regulatory T cells (Treg) and effector T helper cells (Th1, Th2, and Th17). (B): In dogs with gut dysbiosis and secondary gut inflammation, several events contribute to increased bacterial exposure, including mucus layer disruption, dysregulation of epithelial tight junctions, increased intestinal permeability, and increased bacterial adherence to epithelial cells. TLRs initiate the pro-inflammatory stimuli promoting innate local immunity through the recognition of pathogen-associated molecular patterns (PAMPs), present in bacterial antigens, such as lipopolysaccharides (LPS), peptidoglycans, flagella or unmethylated bacterial DNA CpG motifs [72]. The contact of TLRs with PAMPs initiate the innate immune response leading to secretion of cytokines and chemokines and increased expression of adhesion molecules that stimulate and facilitate specialized cells migration responsible for triggering the innate and, subsequently, the adaptive immune response (tumor necrosis factor α (TNF-α), IL-1β, IL-6, IL-12, IL-23, and chemokines) [73]. PAMPs also induce dendritic cells (DCs) maturation that travel to mesenteric lymph nodes and present antigen to naive T cells, which differentiate into Treg and Th17 cells [68]. Tregs also contribute to intestinal homeostasis through the production of immunosuppressive cytokines, such as IL-10. Th17 cells are critical in protecting against bacterial infections because stimulates epithelial cells to secrete anti-microbial proteins and recruit neutrophils from the circulation to the gut microenvironment, resulting in a cycle of inflammation (adapted from Abraham & Cho, 2009 [68]).

In many cases, cancer development is correlated with an inflammatory host response directly to the pathogen (e.g., *Helicobacter pylori* and gastric adenocarcinoma) [74]. But in some cases, cancer progression may be linked to 'sterile' inflammatory causes that are not directly associated with infectious agents, but arising from a response to chronic uncontrolled inflammatory irritation and tissue damage, which adds to the malignant transformation [75]. In these cases, the modulatory roles in cancer development and progression are attributed to commensal or pathogenic agents. It has recently become apparent that commensal community members of microorganisms are crucially involved in tumor-promoting inflammation, which a dysbiotic state stimulates pro-inflammatory properties in the intestinal mucosa [75]. One example is intestinal lymphomagenesis associated to gut microorganism changes in host immune and inflammatory responses affecting lymphocytes [62, 66, 76–78].

The first study for microbiota-induced inflammatory tumorigenesis demonstrated that MyD88-dependent signaling controls the expression of several key modifier genes of intestinal tumorigenesis and has a critical role in cancer progression in mouse model of spontaneous intestinal tumorigenesis and in mice treated with multiple injections of azoxymethane [79]. This revealed that innate immune signaling pathway to intestinal microorganisms is an important factor in intestinal tumorigenesis [79].

It was revealed that mucosal associated invariant T (MAIT) cells from human breast ducts mediate a selective T-helper 17 cell response to human breast carcinoma cells exposed to microbial compounds [80]. This result shows that the presence of bacteria in neoplastic epithelial cells can shape the MAIT cells responses by inflammatory mediators during breast carcinogenesis [80]. Using a mouse model of cutaneous T cell lymphoma (CTCL), it was demonstrated that T cell receptor engagement is critical for the T lymphocytes malignant transformation and that disease progression is also dependent on microbiota [81].

Studies emphasize that inflammatory response to microbial commensal does not occur only in sites of direct contact between the tumor and the microbiota. It was demonstrated an increase of intestinal bacteria translocation associated to inflammation and fibrosis in human chronic liver diseases and also TLR4 activation in non hematopoietic cells in liver carcinogenesis [82]. Thus, there is evidence that intestinal microbiota can affect not only local immunity, but also systemic immune responses.

In veterinary medicine, fecal microbial communities analysis revealed significant lower bacterial diversity and distinct microbial communities in dogs with idiopathic inflammatory bowel disease (IBD) compared to healthy control dogs [30, 83]. This intestinal dysbiosis was correlated with an increase in *E. coli*, a group of particular interest due to its ability to stimulate inflammatory cytokines in human and canine patients with IBD [84–86]. In dogs, the fecal microbiota of patients with intestinal lymphoma, multicentric lymphoma and colorectal tumors showed a significant difference in its composition when compared to clinically healthy dogs microbiota [12, 13, 30]. However, whether these described dysbiotic states play a role in carcinogenesis remains to be determined in dogs.

3.4 Metabolic reprogramming

The metabolome is considered the link between genotypes and phenotypes [87]. It constitutes a set of metabolites synthesized by a biological system, which can be identified by recent "omics" technology called metabolomics, that allows the detection, identification and quantification of intermediate metabolism and, therefore, it can better reflect biological changes in tumorigenesis [88].

Oscillations in microbiota composition induce metabolic changes that can result in host phenotype modifications [89]. Bacterial metabolites production is one of the main signaling pathways between host and its microbiome, and metabolic reprogramming is a central feature of cancer, enabling cells to generate more energy and macromolecules for cancer cell growth, proliferation and division [90].

Microbiome-to-host crosstalk occurs by secreting bacterial metabolites and, after absorption, they enter the circulation and reach the target cells, where they exert their biological effects [91]. Microbial metabolites are detected in peripheral blood (blood metabolome) and feces (fecal metabolome) and have been identified as biomarkers of several diseases, including cancer [92, 93]. The interaction between gut microbiome and fecal and blood metabolome include several mechanisms: a) microbiome can affect gut barrier integrity and alter metabolites absorption (in this case, the same metabolite is associated with a species/pathway in the blood and feces, but the effects directions are opposite); b) direct microbiome-host cell interaction results in host systemic modulation (in this case, the species are associated with blood metabolites, but not fecal metabolites) [94]. In a metagenomic and metabolomic study of 1,004 twins, metabolic pathways were associated with 34% of blood and 95% of fecal metabolites and it was estimated that microbiome was involved in a dialog between 71% of feces and 15% of blood metabolites, highlighting the interaction importance between microbiome and systemic and fecal metabolic environments to identify therapeutic and diagnostic targets [94].

Microbiomes of healthy subjects may share similarities in their metabolic pathways and the fecal metabolome provides a functional readout of microbial activity and can be used as an intermediate phenotype mediating host-microbiome interaction [29, 30, 95]. Zierer and associates (2018) showed that fecal metabolome largely reflects gut microbial composition and fecal metabolic profiling thus is a novel tool to explore links among microbiome composition, host phenotypes, and heritable complex traits [96].

To facilitate understanding, the most investigated bacterial metabolites and enzyme activities can be divided according to the expected effects into more protective or harmful to gut health and carcinogenesis [97].

3.4.1 Protective metabolites (tumor-suppressive metabolites)

3.4.1.1 Short chain fatty acids (SCFAs)

Fermentation of non-digestible carbohydrates from dietary fiber generates SCFAs, such as acetate, butyrate, formate, lactate and propionate [98]. The SCFAs have a key role in gut homeostasis maintenance and epithelial integrity including anti-inflammatory and antiproliferative tumor suppressive effects [98]. Butyrate has been correlated with defense against colon and liver cancer, through its well-known role in regulating inflammation and autophagy [99]. Butyrate production is associated to some Firmicutes, *Eubacterium rectale*, *Roseburia spp.*, *Eubacterium hallii, Coprococcus catus, Faecalibacterium prausnitzii* [100]. It is rapidly adsorbed from gut lumen and is preferentially used as an energy source by gut epithelial cells, then its concentration in the systemic circulation is low. Butyrate is fundamental in epigenetic control; once located inside the cell, inhibits activity of histone deacetylases (HDACs) in colonocytes and immune cells, which promotes the hyperacetylation of histones, allowing transcription factors to bind to DNA and genes to be expressed [99]. This has multiple consequences for gene expression and cellular differentiation including: downregulation of pro-inflammatory cytokines (IL-6 and IL-12) in colonic macrophages; induction of differentiation of Treg cells that express transcription factor FOXP3 (crucial role in controlling intestinal inflammation);

and increased acetylation results in higher expression of FOXP3 [99, 101]. As a consequence of HDAC inhibition, butyrate triggers the factor activator protein 1 (AP-1) signaling pathway in the epithelial cell lines that controls cell proliferation and apoptosis [102] (**Figure 3**).

SCFAs modulate several cancer hallmarks, such as cell proliferation, apoptosis and level of expression of certain genes (via inhibition of HDACs), mechanisms that lead to high anticancer activity (**Figure 3**). This protection can affect both stroma and cancer cells, since they have free fatty acid receptors. It was demonstrated that microbial fermentation of high-fiber diet increased concentrations of butyrate in blood and tumor and significantly decreased tumor growth in mouse with lymphoma, suggesting that dietary fiber protects against human lymphoma cancer [104]. A metabolomics-proteomics approach in colorectal cancer provided a mechanistic link between the M2 isoform of a pyruvate kinase (a direct binding target of butyrate) and metabolic remodeling and the antitumorigenic function of butyrate, highlighting an applicable approach to uncovering protein targets for small molecules with biological functions [105].

Studies in veterinary medicine are very scarce. There is one comparative study reporting higher concentrations of β-hydroxybutyrate in blood from dogs with lymphoma than in healthy dogs, but further investigations are essential to understand the significance of this increase [106]. Another research demonstrated that fecal dysbiosis in dogs with acute diarrhea was associated with altered systemic metabolic states, in which concentrations of fecal propionic acid were significantly decreased compared to healthy dogs [107]. In addition, dogs with inflammatory colorectal polyps (ICRP) showed lower amounts of propionic acid and lower proportions of Bifidobacterium compared to feces of control dogs suggesting that the association

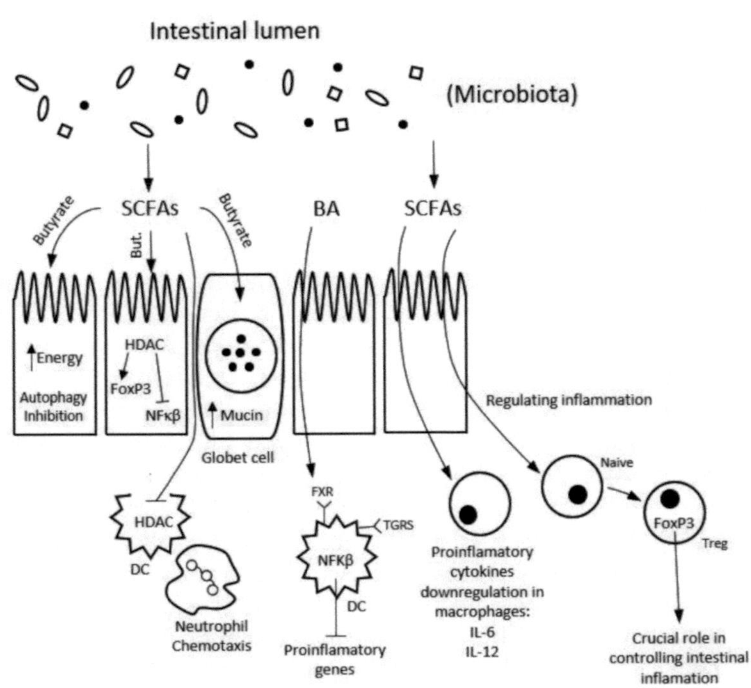

Figure 3.
Modulation of immune signaling through microbial metabolites SCFAs and BA. The metabolic effects directly stimulate the cells of the immune system or are relayed by the intestinal epithelium (adapted from Levy et al., 2019 [103].

between fecal dysbiosis and fecal SCFA concentrations may contribute to ICRP pathogenesis and therapy [108].

3.4.1.2 Phytochemicals

Phytochemicals are bioactive non-nutrient chemical compounds found in fruits, vegetables, grains, and other plant foods, which have biological effects associated with reduced risk of diseases, including cancer [109]. They can be categorized into polyphenols, organosulfur compounds, carotenoids, alkaloids, and nitrogen compounds, but the polyphenols are the most studied ones [109].

Their anti-cancer role includes antioxidant effects, modulation of xenobiotic detoxification pathways and cell proliferation, apoptosis and inflammation [110]. They neutralize reactive oxygen species (ROS) that can damage DNA and predispose to carcinogenesis [97]. A study in human breast cancer cell lines, showed that aqueous extract of the *Pouteria sapota* leaf is rich in phytochemicals with antioxidant properties and significant anti-cancer effects [111]. There is still need for more research and clinical trials in humans and dogs that identify and illustrate the action of phytochemicals.

3.4.2 Harmful metabolite (oncometabolite)

3.4.2.1 Bile acids (BAs)

"Primary" bile acids are synthesized from cholesterol in the liver as cholic acid (CA) and chenodeoxycholic acid (CDCA). When the gallbladder is stimulated after a meal, BA flows into the duodenum and proceeds to the ileum to be actively reabsorbed, returning back to the liver through the portal bloodstream [112]. About 15% of BAs will escape ileum absorption and enter the colon, where the resident microbiota will transform them into secondary BAs (deoxycholic acid, DCA and lithocholic acid, LCA) that have pro and anticancer activity [112]. The enzyme responsible for this conversion is 7α/β hydroxysteroid dehydrogenase (HSDH), and it is produced specially by gram-positive Clostridium species such as *Clostridium scindens* [113].

Quantitative or qualitative BA pool perturbations may greatly affect several BA physiological body functions [113]. The consumption of a high-fat diet changes the gut microbiome and increases the level of DCA, that can promote carcinogenesis in colorectal and liver cancer [114, 115]. Pathways linking BAs to carcinogenesis involve the generation of ROS and reactive nitrogen species (RNS), which cause DNA damage, apoptose and epigenetic changes [112]. Moreover, BAs also exert strong antimicrobial activities, as they damage bacterial cell membranes, contributing to changes in gut microbiota (**Figure 3**). These mechanisms can also be secondary to environmental stimuli (particularly in the context of obesity) and their relationships with human cancer have been recognized as critical in gastrointestinal tract, prostate and breast tissues [116–118].

There are some publications covering changes in the fecal BA profile in canine chronic inflammatory enteropathy and extrahepatic congenital portosystemic shunts, but not in carcinogenesis [119–121].

4. Gut microbiome and therapeutic application

The growing understanding of the microbiome's role in carcinogenesis has allowed the microbiome influence to be linked to the effectiveness of cancer therapies. Microbiome modulation strategies can affect cancer treatment through inactivation or activation of chemotherapeutic agents, modification of immune responses and interference with side effects [72]. This relationship is bilateral, in which the systemic cancer therapy influences gut microbiota, and gut microbiota influences cancer treatment [122]. Recent publications indicate the gut microbiome manipulation as a new treatment tool or to improve the response to cancer therapy. Some of the proposed mechanisms will be discussed below.

4.1 Roles of microbiome in cancer therapy

4.1.1 Chemotherapy

Iida et al. (2013) demonstrated that microbiota impairs disruption response of subcutaneous tumors to platinum derived chemotherapeutic agents. Tumor-bearing mice that lacked microbiota showed therapy efficacy reduction, given that microbiota was important for activating the innate immune response [123]. In another study, administration of *Ruminococcus gnavus* (bacterial strain depleted by treatment with cisplatin) was able to partially restore intestinal mucosa integrity and reduce systemic inflammation in mice treated with cisplatin [124]. Results indicate that reconstitution of gut microbiome can help healing intestinal epithelium in patients treated with chemotherapy.

On the other hand, Viaud et al. (2013) demonstrated that gut microbiota helps shape anti-cancer immune response of cyclophosphamide (CTX). Using mouse models, it was demonstrated that cyclophosphamide alters intestine microbiota composition and induces translocation of selected species into secondary lymphoid organs, resulting in Th17 cells maturation promoting an adaptive immune response against tumors [125]. Daillere et al. (2016) identified *Enterococcus hirae* and *Barnesiella intestinihominis* species involved in tumor immunosurveillance during cyclophosphamide therapy; *E. hirae* translocates from gut to lymph nodes inducing Th1 and Th17 responses mandatory for anti-tumor activity of CTX, while *B. intestinihominis* increases systemic Th1 and CD8 + cytotoxic T cells, which were associated with an increase of IFN-y-producing γ δ tumor infiltrating-lymphocytes (TILs) contributing also for anti-tumor CTX effect [126]. Therefore, cyclophosphamide immunomodulatory effects require a functional microbiome.

Chemotherapy efficacy can also be impacted by intratumoral bacteria. Geller et al. (2017) showed, in a colon cancer mouse model, that Gammaproteobacteria can metabolize chemotherapeutic gemcitabine into an inactive form inducing chemotherapy resistance and that this effect was reversed by antibiotic ciprofloxacin. Interestingly, about 76% of human pancreatic ductal adenocarcinomas were positive for bacteria, mainly Gammaproteobacteria [127]. Perhaps the treatment for this tumor type may be improved by adding antibiotics to the chemotherapy.

Chemotherapy-induced diarrhea (CD) is a frequent adverse event in dogs, in which changes in gut microbiota appear to play a key role. A recent study of 60 dogs undergoing chemotherapy supported the administration of smectite, a natural medical clay, widely used in acute diarrhea treatment in humans, as a first-line treatment of CD in dogs. Interestingly, smectite has anti-inflammatory properties to decrease intestinal bacterial translocation and stabilize intestinal microbiome [128]. However, studies associating microbiome and effectiveness of chemotherapy are still scarce in veterinary medicine.

4.1.2 Immunotherapy

Several studies have shown a complex crosstalk between bacteria and immune host response in the anti-tumor battle. For example, Paulos et al. (2007) reported that total body irradiated mice showed a more efficient anti-tumor response to adoptively transferred tumor-specific CD8+ T cells against melanoma after gut microbial translocation to mesenteric lymph nodes. They observed that the radiation induced the release of microbial LPS and activated innate immune response by TLR4 stimulation and then increased anti-tumor CD8+ T cells, while reduction of host microflora using antibiotics, neutralization of serum LPS using polymyxin B, or removal of LPS signaling were associated with a decrease of anti-tumor response [129].

There is also evidence that gut microbiome modulates efficacy of immune checkpoint inhibitors (CIs), that are monoclonal antibodies with inhibitory effect to specific receptors on T cells and tumor cells, blocking signaling pathways that negatively modulate immune system, allowing specific T cells to promote destruction of cancer cells [130]. Those receptors include cytotoxic T lymphocyte-associated antigen 4 (CTLA-4), programmed death 1 protein (PD-1), and programmed death-ligand 1 (PD-L1) [131].

It was demonstrated that oral administration of Bifidobacterium in mice with melanoma was associated with the same degree of antitumor effects as in those mice that received therapy with PD-L1 antibodies, and the combination of both treatments almost abolished tumor growth [132]. In addition, a report of human gut microbiome metagenomic profiling in 39 metastatic melanoma patients treated with anti-PD1 and/or anti-CTLA-4 immunotherapy identified that those who respond to all types of CIs were enriched for *Bacteroides caccae*, enhancing that microbiota may modulate cancer immunotherapy [133]. These correlations have not yet been demonstrated in dogs.

4.2 Therapeutic manipulation of microbiome and its relevance to cancer therapies response

Given the increasing evidence on the significant role that microbiome can play in cancer, microbiota modulation represents a new therapeutic potential capable of altering disease development. These therapies aim to change the microbial community associated with dysbiosis for those associated with health. In small animals, microbiome manipulations are often described as part of gastrointestinal diseases treatment [134]. Mainly as an adjuvant treatment for cancer, these interventions and their effectiveness are not well established and have only recently been described in literature. The following will discuss some ways in which microbiome can be modified:

4.2.1 Prebiotics and symbiotics

Prebiotics are specific chemicals, capable of promoting growth of a selective group of bacteria and their specific metabolites and thus modulating microbiota in a beneficial way, which may help on anti-tumor treatment [135]. These are non-digestible or absorbable dietary fibers and include fructans (oligofructose and inulin), nonstarch polysaccharides found in some cereal grains, algae, disaccharides (lactulose), and polysaccharides including fructooligosaccharides (FOS) [22].

According to Villegér and colleagues, the effect of prebiotics depends on the presence of beneficial bacteria in the host's intestines [33]. Thus, the combination of probiotics and prebiotics, known as symbiotic, looks promising. Dietary treatment with inulin or oligofructose has been demonstrated to selectively stimulate growth of specific bacterial taxa and alter SCFA levels within the gut [136]. Moreover, these prebiotics reduced the incidence of mammary tumors in rats, significantly potentiated chemotherapy effects as well as RT [136]. The perioperative administration of symbiotics, probiotics (strains Lactobacillus and Bifidobacterium) and prebiotic (fructooligosaccharides), reduced postoperative mortality and complication rates in cancer patients undergoing surgery [137].

The effects of prebiotics were evaluated in dogs, but without focusing on the benefits of cancer treatment. A recent study evaluated the effects of prebiotics in different concentrations in healthy adult dogs and concluded that the galactooligosaccharide prebiotic at 1.0% improved the immunity of healthy dogs [138]. Inulin intervention resulted in a modulation of intestinal bacteria, increase of fecal SCFA and BA in dogs. Given that some studies showed similar dysbiotic states between dogs and humans with cancer [12, 13, 37], it seems relevant that the new approaches to increase anticancer therapy efficiency should include the potential benefits of prebiotic supplementation for both dogs and humans.

4.2.2 Probiotics

Probiotics refers to live bacteria that can be orally administered and confer health beneficial when delivered in adequate amounts [139]. Probiotics colonize the gut temporarily and act modifying colonic environment. Different mechanisms are involved in probiotics protective role: increase in barrier function, epithelial tight junctions integrity, immune response modulation, anti-inflammatory cytokines production, pathogenic bacteria growth inhibition by antimicrobial and antitoxin compound production (i.e. SCFA), and production of enzymatic activities and/or beneficial metabolites to the host [140]. A recent systematic review and meta-analysis investigated probiotics efficacy and safety in patients diagnosed with cancer and concluded that probiotics may be beneficial but further studies are still required [141].

The strains of Lactobacillus and Bifidobacterium are most frequently reported in studies with probiotics. The "protective" effect against colorectal cancer was demonstrated after oral supplement containing *Lactobacillus helveticus* in mice with colonic cancer, in which tumor growth rate and degree of hyperplasia were reduced [142]. These effects were secondary to suppression of NF-κB, increased of anti-inflammatory IL-10 and decreased IL-17-producing T cells [142]. In addition, administration of *L. acidophilus* in mice with breast tumors reduced tumor growth due to altered cytokine production and, in a murine melanoma model, the therapy with aerosolized *L. rhamnosus* promoted immunity against lung metastases, identifying a role for a probiotic cancer "preventing" [143, 144].

Probiotics can also affect patient "outcomes". In a prospective randomized study, after transurethral resection of bladder cancer, the group of patients who received oral supplementation with *L. casei* associated with intravesical epirubicin application had a 3-year recurrence-free survival rate significantly higher than in the isolated chemotherapy group [145]. In addition, some studies demonstrate the action of probiotics on treatment-related toxicity. *L. rhamnosus* decreased diarrhea and abdominal discomfort in patients with colon cancer treated with 5-fluorouracil chemotherapy [146]. Symbiotics (a combination of *Bifidobacterium breve* and *L. casei*) during neoadjuvant chemotherapy in esophageal cancer patients reduced the occurrence of adverse events (diarrhea, neutropenia and lymphopenia) [147].

However, caution should be exercised in their use, since the composition of commercially available probiotics have been inadequately studied, as well as their long-term impact on intestinal microbiota and general health [135]. When investigating the use of pre- and probiotics in dogs, scientific evidence of their benefit is scarce, especially in cancer. Furthermore, the knowledge about appropriate doses and compositions is small in companion animals [148]. Studies suggest that they may have beneficial effects on canine IBD. In a prospective randomized study, 34 IBD dogs received prednisone with or without multi-strain probiotic. Both treatments increased the numbers of total bacteria and were associated with rapid clinical remission but not improvement in histopathologic inflammation [149]. A protective effect of multi-strain probiotic (strains of Lactobacillus, Bifidobacterium and Streptococcus) was also observed in dogs with IBD compared with a control group (treated with metronidazole and prednisolone), with a significant decrease in clinical and histological scores [150].

A recent study showed that probiotics consumption (*L. casei*, *L. plantarum* and Bifidobacterium) in healthy dogs of different age groups, significantly increased beneficial intestinal bacteria (Lactobacillus and *Faecalibacterium prausnitzii*) and decreased potentially harmful bacteria (*E. coli* and *Sutterella stercoricanisin*) mostly in elderly dogs, suggesting that probiotic treatment improves host health and immunity [151].

4.2.3 Fecal microbiome transplantation (FMT)

In FMT, feces are transferred from a healthy donor to the intestinal tract of a diseased recipient [30]. FMT may be delivered via colonoscopy, enema or oral administration, with equal clinical efficacy [152]. The beneficial mechanisms of FMT are still unknown. Nowadays, FMT has been used in resistant *Clostridium difficile* treatment with high response rates [153]. Contrary to gastrointestinal (GI) diseases, application of FMT in cancer is still limited and data was obtained mainly in animal models. The reconstitution of germ-free mice with fecal material from patients with melanoma responsive to anti-PD-L1 and to anti-PD-1 therapies led to better tumor control in contrast to those that received faces from unresponsive patients [154, 155].

The use of FMT in veterinary medicine was studied mostly in dogs with GI diseases, such as in parvovirus-infected puppies and patients with diarrhea due to IBD and *C. perfringens*, and it was associated with faster resolution of clinical signs [156]. For a deep learning regarding the FMT effects in veterinary non-oncological diseases and the potential applications of FMT in animals, including therapeutic, prophylactic and immunogenic uses, the reader may consult Niederwerder (2018) publication [157].

5. Conclusions

Host-microbiota interactions are crucial in human and animal health and disease development, yet microbiota function and dynamics during disease states are only partially understood. There is growing evidence supporting that immunoregulatory and anti-inflammatory effects of gut and tumor microbiota are essential in the battle of cancer. However, most studies were performed in preclinical models, which have many pitfalls in regard of spontaneous cancer research urging the need for clinical studies benefiting both species.

Acknowledgements

This manuscript received no specific grant from any funding agency in the public, commercial, or not-for-profit sectors.

Conflict of interest

The authors declare no conflict of interest.

Author details

Tatiane Moreno Ferrarias Epiphanio[1] and Andreia A.F. Santos[1,2*]

1 Department of Veterinary Clinics of the Biomedical Sciences, Institute of Abel Salazar (ICBAS), University of Porto, Porto, Portugal

2 Animal Science and Study Centre/Food and Agrarian Sciences and Technologies, Institute (CECA/ICETA), University of Porto, Porto, Portugal

*Address all correspondence to: aasantos@icbas.up.pt

References

[1] Gopalakrishnan V, Helmink BA, Spencer CN, Reuben A, Wargo JA. The Influence of the Gut Microbiome on Cancer, Immunity, and Cancer Immunotherapy. Cancer Cell. 2018;33(4):570-580. doi:10.1016/j.ccell.2018.03.015

[2] Marchesi JR, Ravel J. The vocabulary of microbiome research: a proposal. Microbiome. 2015;3:31. Published 2015 Jul 30. doi:10.1186/s40168-015-0094-5

[3] Berg G, Rybakova D, Fischer D, et al. Microbiome definition re-visited: old concepts and new challenges [published correction appears in Microbiome. 2020 Aug 20;8(1):119]. Microbiome. 2020;8(1):103. Published 2020 Jun 30. doi:10.1186/s40168-020-00875-0

[4] Deng P, Swanson KS. Gut microbiota of humans, dogs and cats: current knowledge and future opportunities and challenges. Br J Nutr. 2015;113 Suppl:S6-S17. doi:10.1017/S0007114514002943

[5] Sherwani MA, Tufail S, Muzaffar AF, Yusuf N. The skin microbiome and immune system: Potential target for chemoprevention?. Photodermatol Photoimmunol Photomed. 2018;34(1):25-34. doi:10.1111/phpp.12334

[6] Meason-Smith C, Diesel A, Patterson AP, et al. Characterization of the cutaneous mycobiota in healthy and allergic cats using next generation sequencing. Vet Dermatol. 2017;28(1):71-e17. doi:10.1111/vde.12373

[7] Isaiah A, Hoffmann AR, Kelley R, Mundell P, Steiner JM, Suchodolski JS. Characterization of the nasal and oral microbiota of detection dogs. PLoS One. 2017;12(9):e0184899. Published 2017 Sep 21. doi:10.1371/journal.pone.0184899

[8] Meng S, Chen B, Yang J, et al. Study of Microbiomes in Aseptically Collected Samples of Human Breast Tissue Using Needle Biopsy and the Potential Role of in situ Tissue Microbiomes for Promoting Malignancy. Front Oncol. 2018;8:318. Published 2018 Aug 17. doi:10.3389/fonc.2018.00318 Whiteside SA, Razvi H, Dave S, Reid G, Burton JP. The microbiome of the urinary tract--a role beyond infection. Nat Rev Urol. 2015;12(2):81-90. doi:10.1038/nrurol.2014.361

[9] Burton EN, Cohn LA, Reinero CN, Rindt H, Moore SG, Ericsson AC. Characterization of the urinary microbiome in healthy dogs. PLoS One. 2017;12(5):e0177783. Published 2017 May 17. doi:10.1371/journal.pone.0177783

[10] Belkaid Y, Hand TW. Role of the microbiota in immunity and inflammation. Cell. 2014;157(1):121-141. doi:10.1016/j.cell.2014.03.011

[11] Vogtmann E, Goedert JJ. Epidemiologic studies of the human microbiome and cancer. Br J Cancer. 2016;114(3):237-242. doi:10.1038/bjc.2015.465

[12] Gavazza A, Rossi G, Lubas G, Cerquetella M, Minamoto Y, Suchodolski JS. Faecal microbiota in dogs with multicentric lymphoma. Vet Comp Oncol. 2018;16(1):E169-E175. doi:10.1111/vco.12367

[13] Herstad KMV, Moen AEF, Gaby JC, Moe L, Skancke E. Characterization of the fecal and mucosa-associated microbiota in dogs with colorectal epithelial tumors. PLoS One. 2018;13(5):e0198342. Published 2018 May 31. doi:10.1371/journal.pone.0198342

[14] Fleming JM, Creevy KE, Promislow DE. Mortality in north american dogs from 1984 to 2004: an investigation into age-, size-, and breed-related causes of death.

J Vet Intern Med. 2011;25(2):187-198. doi:10.1111/j.1939-1676.2011.0695.x

[15] Hansen K, Khanna C. Spontaneous and genetically engineered animal models; use in preclinical cancer drug development. Eur J Cancer. 2004;40(6):858-880. doi:10.1016/j.ejca.2003.11.031

[16] Bronson RT. Variation in age at death of dogs of different sexes and breeds. Am J Vet Res. 1982;43(11):2057-2059.

[17] Bray F, Ferlay J, Soerjomataram I, Siegel RL, Torre LA, Jemal A. Global cancer statistics 2018: GLOBOCAN estimates of incidence and mortality worldwide for 36 cancers in 185 countries [published correction appears in CA Cancer J Clin. 2020 Jul;70(4):313]. CA Cancer J Clin. 2018;68(6):394-424. doi:10.3322/caac.21492

[18] Paoloni M, Khanna C. Translation of new cancer treatments from pet dogs to humans. Nat Rev Cancer. 2008;8(2):147-156. doi:10.1038/nrc2273

[19] Schiffman JD, Breen M. Comparative oncology: what dogs and other species can teach us about humans with cancer. Philos Trans R Soc Lond B Biol Sci. 2015;370(1673):20140231. doi:10.1098/rstb.2014.0231

[20] Coelho LP, Kultima JR, Costea PI, et al. Similarity of the dog and human gut microbiomes in gene content and response to diet. Microbiome. 2018;6(1):72. Published 2018 Apr 19. doi:10.1186/s40168-018-0450-3

[21] Song SJ, Lauber C, Costello EK, et al. Cohabiting family members share microbiota with one another and with their dogs. Elife. 2013;2:e00458. Published 2013 Apr 16. doi:10.7554/eLife.00458

[22] Barko PC, McMichael MA, Swanson KS, Williams DA. The Gastrointestinal Microbiome: A Review. J Vet Intern Med. 2018;32(1):9-25. doi:10.1111/jvim.14875

[23] Mondo E, Marliani G, Accorsi PA, Cocchi M, Di Leone A. Role of gut microbiota in dog and cat's health and diseases. Open Vet J. 2019;9(3):253-258. doi:10.4314/ovj.v9i3.10

[24] Dominguez-Bello MG, Godoy-Vitorino F, Knight R, Blaser MJ. Role of the microbiome in human development. Gut. 2019;68(6):1108-1114. doi:10.1136/gutjnl-2018-317503

[25] Scott AJ, Alexander JL, Merrifield CA, et al. International Cancer Microbiome Consortium consensus statement on the role of the human microbiome in carcinogenesis. Gut. 2019;68(9):1624-1632. doi:10.1136/gutjnl-2019-318556

[26] Young VB. The role of the microbiome in human health and disease: an introduction for clinicians. BMJ. 2017;356:j831. Published 2017 Mar 15. doi:10.1136/bmj.j831

[27] Turnbaugh PJ, Ley RE, Hamady M, Fraser-Liggett CM, Knight R, Gordon JI. The human microbiome project. Nature. 2007;449(7164):804-810. doi:10.1038/nature06244

[28] Deng P, Swanson KS. Gut microbiota of humans, dogs and cats: current knowledge and future opportunities and challenges. Br J Nutr. 2015;113 Suppl:S6-S17. doi:10.1017/S0007114514002943

[29] Collins D, Hogan AM, Winter DC. Microbial and viral pathogens in colorectal cancer. Lancet Oncol. 2011;12(5):504-512. doi:10.1016/S1470-2045(10)70186-8

[30] Omori M, Maeda S, Igarashi H, et al. Fecal microbiome in dogs with inflammatory bowel disease and intestinal lymphoma. J Vet Med Sci.

2017;79(11):1840-1847. doi:10.1292/jvms.17-0045

[31] Arthur JC, Perez-Chanona E, Mühlbauer M, et al. Intestinal inflammation targets cancer-inducing activity of the microbiota. Science. 2012;338(6103):120-123. doi:10.1126/science.1224820

[32] Schwabe RF, Jobin C. The microbiome and cancer. Nat Rev Cancer. 2013;13(11):800-812. doi:10.1038/nrc3610

[33] Villéger R, Lopès A, Carrier G, et al. Intestinal Microbiota: A Novel Target to Improve Anti-Tumor Treatment?. Int J Mol Sci. 2019;20(18):4584. Published 2019 Sep 17. doi:10.3390/ijms20184584

[34] Grice EA, Segre JA. The skin microbiome [published correction appears in Nat Rev Microbiol. 2011 Aug;9(8):626]. Nat Rev Microbiol. 2011;9(4):244-253. doi:10.1038/nrmicro2537

[35] Buchta V. Vaginal microbiome. Vaginální mikrobiom. Ceska Gynekol. 2018;83(5):371-379.

[36] Rea K, Dinan TG, Cryan JF. Gut Microbiota: A Perspective for Psychiatrists. Neuropsychobiology. 2020;79(1):50-62. doi:10.1159/000504495

[37] Picardo SL, Coburn B, Hansen AR. The microbiome and cancer for clinicians. Crit Rev Oncol Hematol. 2019;141:1-12. doi:10.1016/j.critrevonc.2019.06.004

[38] Hanahan D, Weinberg RA. Hallmarks of cancer: the next generation. Cell. 2011;144(5):646-674. doi:10.1016/j.cell.2011.02.013

[39] Münger K, Baldwin A, Edwards KM, et al. Mechanisms of human papillomavirus-induced oncogenesis. J Virol. 2004;78(21):11451-11460. doi:10.1128/JVI.78.21.11451-11460.2004

[40] Pagano JS, Blaser M, Buendia MA, et al. Infectious agents and cancer: criteria for a causal relation. Semin Cancer Biol. 2004;14(6):453-471. doi:10.1016/j.semcancer.2004.06.009

[41] Morales-Sánchez A, Fuentes-Pananá EM. Human viruses and cancer. Viruses. 2014;6(10):4047-4079. Published 2014 Oct 23. doi:10.3390/v6104047

[42] Thaiwong T, Sledge DG, Wise AG, Olstad K, Maes RK, Kiupel M. Malignant transformation of canine oral papillomavirus (CPV1)-associated papillomas in dogs: An emerging concern?. Papillomavirus Res. 2018;6:83-89. doi:10.1016/j.pvr.2018.10.007

[43] Munday JS, Thomson NA, Luff JA. Papillomaviruses in dogs and cats. Vet J. 2017;225:23-31. doi:10.1016/j.tvjl.2017.04.018

[44] Regalado Ibarra AM, Legendre L, Munday JS. Malignant Transformation of a Canine Papillomavirus Type 1-Induced Persistent Oral Papilloma in a 3-Year-Old Dog. J Vet Dent. 2018;35(2):79-95. doi:10.1177/0898756418774575

[45] Munday JS, O'Connor KI, Smits B. Development of multiple pigmented viral plaques and squamous cell carcinomas in a dog infected by a novel papillomavirus. Vet Dermatol. 2011;22(1):104-110. doi:10.1111/j.1365-3164.2010.00913.x

[46] Chang CY, Chen WT, Haga T, et al. The Detection and Association of Canine Papillomavirus with Benign and Malignant Skin Lesions in Dogs. Viruses. 2020;12(2):170. Published 2020 Feb 3. doi:10.3390/v12020170

[47] Luff J, Mader M, Rowland P, Britton M, Fass J, Yuan H. Viral genome

integration of canine papillomavirus 16. Papillomavirus Res. 2019;7:88-96. doi:10.1016/j.pvr.2019.02.002

[48] Candela M, Guidotti M, Fabbri A, Brigidi P, Franceschi C, Fiorentini C. Human intestinal microbiota: cross-talk with the host and its potential role in colorectal cancer. Crit Rev Microbiol. 2011;37(1):1-14. doi:10.3109/1040841X.2010.501760

[49] Rosadi F, Fiorentini C, Fabbri A. Bacterial protein toxins in human cancers. Pathog Dis. 2016;74(1):ftv105. doi:10.1093/femspd/ftv105

[50] Guerra L, Cortes-Bratti X, Guidi R, Frisan T. The biology of the cytolethal distending toxins. Toxins (Basel). 2011;3(3):172-190. doi:10.3390/toxins3030172

[51] Nougayrède JP, Homburg S, Taieb F, et al. *Escherichia coli* induces DNA double-strand breaks in eukaryotic cells. Science. 2006;313(5788):848-851. doi:10.1126/science.1127059

[52] Cortes-Bratti X, Karlsson C, Lagergård T, Thelestam M, Frisan T. The Haemophilus ducreyi cytolethal distending toxin induces cell cycle arrest and apoptosis via the DNA damage checkpoint pathways. J Biol Chem. 2001;276(7):5296-5302. doi:10.1074/jbc.M008527200

[53] Faïs T, Delmas J, Serres A, Bonnet R, Dalmasso G. Impact of CDT Toxin on Human Diseases. Toxins (Basel). 2016;8(7):220. Published 2016 Jul 15. doi:10.3390/toxins8070220

[54] Buc E, Dubois D, Sauvanet P, et al. High prevalence of mucosa-associated *E. coli* producing cyclomodulin and genotoxin in colon cancer. PLoS One. 2013;8(2):e56964. doi:10.1371/journal.pone.0056964

[55] Dalmasso G, Cougnoux A, Delmas J, Darfeuille-Michaud A, Bonnet R. The bacterial genotoxin colibactin promotes colon tumor growth by modifying the tumor microenvironment. Gut Microbes. 2014;5(5):675-680. doi:10.4161/19490976.2014.969989

[56] Wilson MR, Jiang Y, Villalta PW, et al. The human gut bacterial genotoxin colibactin alkylates DNA. Science. 2019;363(6428):eaar7785. doi:10.1126/science.aar7785

[57] Cougnoux A, Dalmasso G, Martinez R, et al. Bacterial genotoxin colibactin promotes colon tumour growth by inducing a senescence-associated secretory phenotype. Gut. 2014;63(12):1932-1942. doi:10.1136/gutjnl-2013-305257

[58] Feng Y, Mannion A, Madden CM, et al. Cytotoxic *Escherichia coli* strains encoding colibactin and cytotoxic necrotizing factor (CNF) colonize laboratory macaques. Gut Pathog. 2017;9:71. Published 2017 Dec 6. doi:10.1186/s13099-017-0220-y

[59] Kurilshikov A, Wijmenga C, Fu J, Zhernakova A. Host Genetics and Gut Microbiome: Challenges and Perspectives. Trends Immunol. 2017;38(9):633-647. doi:10.1016/j.it.2017.06.003

[60] Cebra JJ. Influences of microbiota on intestinal immune system development. Am J Clin Nutr. 1999;69(5):1046S–1051S. doi:10.1093/ajcn/69.5.1046s

[61] Thaiss CA, Zmora N, Levy M, Elinav E. The microbiome and innate immunity. Nature. 2016;535(7610):65-74. doi:10.1038/nature18847

[62] Honda K, Littman DR. The microbiota in adaptive immune homeostasis and disease. Nature. 2016;535(7610):75-84. doi:10.1038/nature18848

[63] Sonnenberg GF, Monticelli LA, Alenghat T, et al. Innate lymphoid cells

promote anatomical containment of lymphoid-resident commensal bacteria. Science. 2012;336(6086):1321-1325. doi:10.1126/science.1222551

[64] Zheng D, Liwinski T, Elinav E. Interaction between microbiota and immunity in health and disease. Cell Res. 2020;30(6):492-506. doi:10.1038/s41422-020-0332-7

[65] Festi D, Schiumerini R, Birtolo C, et al. Gut microbiota and its pathophysiology in disease paradigms. Dig Dis. 2011;29(6):518-524. doi:10.1159/000332975

[66] Huang T, Wei B, Velazquez P, Borneman J, Braun J. Commensal microbiota alter the abundance and TCR responsiveness of splenic naïve CD4+ T lymphocytes. Clin Immunol. 2005;117(3):221-230. doi:10.1016/j.clim.2005.09.012

[67] Khan AA, Shrivastava A, Khurshid M. Normal to cancer microbiome transformation and its implication in cancer diagnosis. Biochim Biophys Acta. 2012;1826(2):331-337. doi:10.1016/j.bbcan.2012.05.005

[68] Abraham C, Cho JH. Inflammatory bowel disease. N Engl J Med. 2009;361(21):2066-2078. doi:10.1056/NEJMra0804647

[69] Kawai T, Akira S. The roles of TLRs, RLRs and NLRs in pathogen recognition. Int Immunol. 2009;21(4):317-337. doi:10.1093/intimm/dxp017

[70] Abt MC, Osborne LC, Monticelli LA, et al. Commensal bacteria calibrate the activation threshold of innate antiviral immunity. Immunity. 2012;37(1):158-170. doi:10.1016/j.immuni.2012.04.011

[71] Frosali S, Pagliari D, Gambassi G, Landolfi R, Pandolfi F, Cianci R. How the Intricate Interaction among Toll-Like Receptors, Microbiota, and Intestinal Immunity Can Influence Gastrointestinal Pathology. J Immunol Res. 2015;2015:489821. doi:10.1155/2015/489821

[72] Polk DB, Peek RM Jr. Helicobacter pylori: gastric cancer and beyond [published correction appears in Nat Rev Cancer. 2010 Aug;10(8):593]. Nat Rev Cancer. 2010;10(6):403-414. doi:10.1038/nrc2857

[73] Elinav E, Nowarski R, Thaiss CA, Hu B, Jin C, Flavell RA. Inflammation-induced cancer: crosstalk between tumours, immune cells and microorganisms. Nat Rev Cancer. 2013;13(11):759-771. doi:10.1038/nrc3611

[74] Wei B, Wingender G, Fujiwara D, et al. Commensal microbiota and CD8+ T cells shape the formation of invariant NKT cells. J Immunol. 2010;184(3):1218-1226. doi:10.4049/jimmunol.0902620

[75] Mazmanian SK, Liu CH, Tzianabos AO, Kasper DL. An immunomodulatory molecule of symbiotic bacteria directs maturation of the host immune system. Cell. 2005;122(1):107-118. doi:10.1016/j.cell.2005.05.007

[76] Yamamoto ML, Schiestl RH. Intestinal microbiome and lymphoma development. Cancer J. 2014;20(3):190-194. doi:10.1097/PPO.0000000000000047

[77] Rakoff-Nahoum S, Medzhitov R. Regulation of spontaneous intestinal tumorigenesis through the adaptor protein MyD88. Science. 2007;317(5834):124-127. doi:10.1126/science.1140488

[78] Zumwalde NA, Haag JD, Gould MN, Gumperz JE. Mucosal associated invariant T cells from human breast ducts mediate a Th17-skewed response to bacterially exposed breast carcinoma

cells. Breast Cancer Res. 2018;20(1):111. Published 2018 Sep 12. doi:10.1186/s13058-018-1036-5

[79] Fanok MH, Sun A, Fogli LK, et al. Role of Dysregulated Cytokine Signaling and Bacterial Triggers in the Pathogenesis of Cutaneous T-Cell Lymphoma. J Invest Dermatol. 2018;138(5):1116-1125. doi:10.1016/j.jid.2017.10.028

[80] Dapito DH, Mencin A, Gwak GY, et al. Promotion of hepatocellular carcinoma by the intestinal microbiota and TLR4. Cancer Cell. 2012;21(4):504-516. doi:10.1016/j.ccr.2012.02.007

[81] Minamoto Y, Otoni CC, Steelman SM, et al. Alteration of the fecal microbiota and serum metabolite profiles in dogs with idiopathic inflammatory bowel disease. Gut Microbes. 2015;6(1):33-47. doi:10.1080/19490976.2014.997612

[82] De la Fuente M, Franchi L, Araya D, et al. *Escherichia coli* isolates from inflammatory bowel diseases patients survive in macrophages and activate NLRP3 inflammasome [published correction appears in Int J Med Microbiol. 2015 May;305(3):434]. Int J Med Microbiol. 2014;304(3-4):384-392. doi:10.1016/j.ijmm.2014.01.002

[83] Vázquez-Baeza Y, Hyde ER, Suchodolski JS, Knight R. Dog and human inflammatory bowel disease rely on overlapping yet distinct dysbiosis networks. Nat Microbiol. 2016;1:16177. Published 2016 Oct 3. doi:10.1038/nmicrobiol.2016.177

[84] Cassmann E, White R, Atherly T, et al. Alterations of the Ileal and Colonic Mucosal Microbiota in Canine Chronic Enteropathies. PLoS One. 2016;11(2):e0147321. Published 2016 Feb 3. doi:10.1371/journal.pone.0147321

[85] Fiehn O. Metabolomics--the link between genotypes and phenotypes. Plant Mol Biol. 2002;48(1-2):155-171.

[86] Holmes E, Wilson ID, Nicholson JK. Metabolic phenotyping in health and disease. Cell. 2008;134(5):714-717. doi:10.1016/j.cell.2008.08.026

[87] Tao J, Li S, Gan RY, Zhao CN, Meng X, Li HB. Targeting gut microbiota with dietary components on cancer: Effects and potential mechanisms of action. Crit Rev Food Sci Nutr. 2020;60(6):1025-1037. doi:10.1080/10408398.2018.1555789

[88] Wang L, Liu X, Yang Q. Application of Metabolomics in Cancer Research: As a Powerful Tool to Screen Biomarker for Diagnosis, Monitoring and Prognosis of Cancer. Biomark J. (2018); 4:12. doi: 10.21767/2472-1646.100050.

[89] Mikó E, Kovács T, Sebő É, et al. Microbiome-Microbial Metabolome-Cancer Cell Interactions in Breast Cancer-Familiar, but Unexplored. Cells. 2019;8(4):293. Published 2019 Mar 29. doi:10.3390/cells8040293

[90] Velmurugan G, Dinakaran V, Rajendhran J, Swaminathan K. Blood Microbiota and Circulating Microbial Metabolites in Diabetes and Cardiovascular Disease. Trends Endocrinol Metab. 2020;31(11):835-847. doi:10.1016/j.tem.2020.01.013

[91] Yang Y, Misra BB, Liang L, et al. Integrated microbiome and metabolome analysis reveals a novel interplay between commensal bacteria and metabolites in colorectal cancer. Theranostics. 2019;9(14):4101-4114. Published 2019 May 31. doi:10.7150/thno.35186

[92] Visconti A, Le Roy CI, Rosa F, et al. Interplay between the human gut microbiome and host metabolism. Nat Commun. 2019;10(1):4505. Published 2019 Oct 3. doi:10.1038/s41467-019-12476-z

[93] Marcobal A, Kashyap PC, Nelson TA, et al. A metabolomic view

of how the human gut microbiota impacts the host metabolome using humanized and gnotobiotic mice. ISME J. 2013;7(10):1933-1943. doi:10.1038/ismej.2013.89Human Microbiome Project Consortium. A framework for human microbiome research. Nature. 2012;486(7402):215-221. Published 2012 Jun 13. doi:10.1038/nature11209

[94] Zierer J, Jackson MA, Kastenmüller G, et al. The fecal metabolome as a functional readout of the gut microbiome. Nat Genet. 2018;50(6):790-795. doi:10.1038/s41588-018-0135-7

[95] Louis P, Hold GL, Flint HJ. The gut microbiota, bacterial metabolites and colorectal cancer. Nat Rev Microbiol. 2014;12(10):661-672. doi:10.1038/nrmicro3344

[96] Morrison DJ, Preston T. Formation of short chain fatty acids by the gut microbiota and their impact on human metabolism. Gut Microbes. 2016;7(3):189-200. doi:10.1080/19490976.2015.1134082

[97] Encarnação JC, Abrantes AM, Pires AS, Botelho MF. Revisit dietary fiber on colorectal cancer: butyrate and its role on prevention and treatment. Cancer Metastasis Rev. 2015;34(3):465-478. doi:10.1007/s10555-015-9578-9

[98] Reichardt N, Duncan SH, Young P, et al. Phylogenetic distribution of three pathways for propionate production within the human gut microbiota [published correction appears in ISME J. 2014 Jun;8(6):1352]. ISME J. 2014;8(6):1323-1335. doi:10.1038/ismej.2014.14

[99] Furusawa Y, Obata Y, Fukuda S, et al. Commensal microbe-derived butyrate induces the differentiation of colonic regulatory T cells [published correction appears in Nature. 2014 Feb 13;506(7487):254]. Nature. 2013;504(7480):446-450. doi:10.1038/nature12721

[100] Nepelska M, Cultrone A, Béguet-Crespel F, et al. Butyrate produced by commensal bacteria potentiates phorbol esters induced AP-1 response in human intestinal epithelial cells. PLoS One. 2012;7(12):e52869. doi:10.1371/journal.pone.0052869

[101] Levy M, Blacher E, Elinav E. Microbiome, metabolites and host immunity. Curr Opin Microbiol. 2017;35:8-15. doi:10.1016/j.mib.2016.10.003

[102] Wei W, Sun W, Yu S, Yang Y, Ai L. Butyrate production from high-fiber diet protects against lymphoma tumor. Leuk Lymphoma. 2016;57(10):2401-2408. doi:10.3109/10428194.2016.1144879

[103] Li Q, Cao L, Tian Y, et al. Butyrate Suppresses the Proliferation of Colorectal Cancer Cells via Targeting Pyruvate Kinase M2 and Metabolic Reprogramming. Mol Cell Proteomics. 2018;17(8):1531-1545. doi:10.1074/mcp.RA118.000752 McQuown B, Burgess KE, Heinze CR. Preliminary investigation of blood concentrations of insulin-like growth factor, insulin, lactate and β-hydroxybutyrate in dogs with lymphoma as compared with matched controls. Vet Comp Oncol. 2018;16(2):262-267. doi:10.1111/vco.12376

[104] McQuown B, Burgess KE, Heinze CR. Preliminary investigation of blood concentrations of insulin-like growth factor, insulin, lactate and β-hydroxybutyrate in dogs with lymphoma as compared with matched controls. Vet Comp Oncol. 2018;16(2):262-267. doi:10.1111/vco.12376

[105] Guard BC, Barr JW, Reddivari L, et al. Characterization of microbial dysbiosis and metabolomic changes in dogs with acute diarrhea. PLoS One. 2015;10(5):e0127259. Published 2015 May 22. doi:10.1371/journal.pone.0127259

[106] Igarashi H, Ohno K, Matsuki N, et al. Analysis of fecal short chain fatty acid concentration in miniature dachshunds with inflammatory colorectal polyps. J Vet Med Sci. 2017;79(10):1727-1734. doi:10.1292/jvms.17-0165

[107] Lampe JW, Chang JL. Interindividual differences in phytochemical metabolism and disposition. Semin Cancer Biol. 2007;17(5):347-353. doi:10.1016/j.semcancer.2007.05.003

[108] Ramos S. Cancer chemoprevention and chemotherapy: dietary polyphenols and signalling pathways. Mol Nutr Food Res. 2008;52(5):507-526. doi:10.1002/mnfr.200700326

[109] Prabhu DS, Selvam AP, Rajeswari VD. Effective anti-cancer property of *Pouteria sapota* leaf on breast cancer cell lines. Biochem Biophys Rep. 2018;15:39-44. Published 2018 Jun 28. doi:10.1016/j.bbrep.2018.06.004

[110] Di Ciaula A, Wang DQ, Molina-Molina E, et al. Bile Acids and Cancer: Direct and Environmental-Dependent Effects. Ann Hepatol. 2017;16(Suppl. 1: s3-105.):s87-s105. doi:10.5604/01.3001.0010.5501

[111] Dart A. Gut microbiota bile acid metabolism controls cancer immunosurveillance. Nat Rev Microbiol. 2018;16(8):453. doi:10.1038/s41579-018-0053-9

[112] Ou J, DeLany JP, Zhang M, Sharma S, O'Keefe SJ. Association between low colonic short-chain fatty acids and high bile acids in high colon cancer risk populations. Nutr Cancer. 2012;64(1):34-40. doi:10.1080/01635581.2012.630164

[113] Yoshimoto S, Loo TM, Atarashi K, et al. Obesity-induced gut microbial metabolite promotes liver cancer through senescence secretome [published correction appears in Nature. 2014 Feb 20;506(7488):396. Hattori, Masahisa [corrected to Hattori, Masahira]]. Nature. 2013;499(7456):97-101. doi:10.1038/nature12347

[114] Gadaleta RM, Garcia-Irigoyen O, Moschetta A. Bile acids and colon cancer: Is FXR the solution of the conundrum?. Mol Aspects Med. 2017;56:66-74. doi:10.1016/j.mam.2017.04.002

[115] Giaginis C, Karandrea D, Alexandrou P, et al. High Farnesoid X Receptor (FXR) expression is a strong and independent prognosticator in invasive breast carcinoma. Neoplasma. 2017;64(4):633-639. doi:10.4149/neo_2017_420

[116] Goldberg AA, Titorenko VI, Beach A, Sanderson JT. Bile acids induce apoptosis selectively in androgen-dependent and -independent prostate cancer cells. PeerJ. 2013;1:e122. Published 2013 Aug 8. doi:10.7717/peerj.122

[117] Herstad KMV, Rønning HT, Bakke AM, Moe L, Skancke E. Changes in the faecal bile acid profile in dogs fed dry food vs high content of beef: a pilot study. Acta Vet Scand. 2018;60(1):29. Published 2018 May 11. doi:10.1186/s13028-018-0383-7

[118] Guard BC, Honneffer JB, Jergens AE, et al. Longitudinal assessment of microbial dysbiosis, fecal unconjugated bile acid concentrations, and disease activity in dogs with steroid-responsive chronic inflammatory enteropathy. J Vet Intern Med. 2019;33(3):1295-1305. doi:10.1111/jvim.15493

[119] Bristow P, Tivers M, Packer R, et al. Long-term serum bile acid concentrations in 51 dogs after complete extrahepatic congenital portosystemic shunt ligation [published correction appears in J Small Anim

Pract. 2018 Jan;59(1):70]. J Small Anim Pract. 2017;58(8):454-460. doi:10.1111/jsap.12685

[120] Kovács T, Mikó E, Ujlaki G, Sári Z, Bai P. The Microbiome as a Component of the Tumor Microenvironment. Adv Exp Med Biol. 2020;1225:137-153. doi:10.1007/978-3-030-35727-6_10

[121] Aarnoutse R, Ziemons J, Penders J, Rensen SS, de Vos-Geelen J, Smidt ML. The Clinical Link between Human Intestinal Microbiota and Systemic Cancer Therapy. Int J Mol Sci. 2019;20(17):4145. Published 2019 Aug 25. doi:10.3390/ijms20174145

[122] Iida N, Dzutsev A, Stewart CA, et al. Commensal bacteria control cancer response to therapy by modulating the tumor microenvironment. Science. 2013;342(6161):967-970. doi:10.1126/science.1240527

[123] Perales-Puchalt A, Perez-Sanz J, Payne KK, et al. Frontline Science: Microbiota reconstitution restores intestinal integrity after cisplatin therapy. J Leukoc Biol. 2018;103(5):799-805. doi:10.1002/JLB.5HI1117-446RR

[124] Viaud S, Saccheri F, Mignot G, et al. The intestinal microbiota modulates the anticancer immune effects of cyclophosphamide. Science. 2013;342(6161):971-976. doi:10.1126/science.1240537

[125] Daillère R, Vétizou M, Waldschmitt N, et al. Enterococcus hirae and Barnesiella intestinihominis Facilitate Cyclophosphamide-Induced Therapeutic Immunomodulatory Effects. Immunity. 2016;45(4):931-943. doi:10.1016/j.immuni.2016.09.009

[126] Geller LT, Barzily-Rokni M, Danino T, et al. Potential role of intratumor bacteria in mediating tumor resistance to the chemotherapeutic drug gemcitabine. Science. 2017;357(6356):1156-1160. doi:10.1126/science.aah5043

[127] Fournier Q, Serra JC, Williams C, Bavcar S. Chemotherapy-induced diarrhoea in dogs and its management with smectite: Results of a monocentric open-label randomized clinical trial [published online ahead of print, 2020 Jun 20]. Vet Comp Oncol. 2020;10.1111/vco.12631. doi:10.1111/vco.12631

[128] Paulos CM, Wrzesinski C, Kaiser A, et al. Microbial translocation augments the function of adoptively transferred self/tumor-specific CD8+ T cells via TLR4 signaling [published correction appears in J Clin Invest. 2007 Oct;117(10):3140]. J Clin Invest. 2007;117(8):2197-2204. doi:10.1172/JCI32205

[129] Yi M, Yu S, Qin S, et al. Gut microbiome modulates efficacy of immune checkpoint inhibitors. J Hematol Oncol. 2018;11(1):47. Published 2018 Mar 27. doi:10.1186/s13045-018-0592-6

[130] Postow MA, Callahan MK, Wolchok JD. Immune Checkpoint Blockade in Cancer Therapy. J Clin Oncol. 2015;33(17):1974-1982. doi:10.1200/JCO.2014.59.4358

[131] Sivan A, Corrales L, Hubert N, et al. Commensal Bifidobacterium promotes antitumor immunity and facilitates anti-PD-L1 efficacy. Science. 2015;350(6264):1084-1089. doi:10.1126/science.aac4255

[132] Frankel AE, Coughlin LA, Kim J, et al. Metagenomic Shotgun Sequencing and Unbiased Metabolomic Profiling Identify Specific Human Gut Microbiota and Metabolites Associated with Immune Checkpoint Therapy Efficacy in Melanoma Patients. Neoplasia. 2017;19(10):848-855. doi:10.1016/j.neo.2017.08.004

[133] Pilla R, Suchodolski JS. The Role of the Canine Gut Microbiome and Metabolome in Health and Gastrointestinal Disease. Front Vet Sci. 2020;6:498. Published 2020 Jan 14. doi:10.3389/fvets.2019.00498

[134] Helmink BA, Khan MAW, Hermann A, Gopalakrishnan V, Wargo JA. The microbiome, cancer, and cancer therapy. Nat Med. 2019;25(3):377-388. doi:10.1038/s41591-019-0377-7

[135] Taper HS, Roberfroid MB. Possible adjuvant cancer therapy by two prebiotics--inulin or oligofructose. In Vivo. 2005;19(1):201-204.

[136] Sommacal HM, Bersch VP, Vitola SP, Osvaldt AB. Perioperative synbiotics decrease postoperative complications in periampullary neoplasms: a randomized, double-blind clinical trial. Nutr Cancer. 2015;67(3):457-462. doi:10.1080/01635581.2015.1004734

[137] Rentas MF, Pedreira RS, Perini MP, et al. Galactoligosaccharide and a prebiotic blend improve colonic health and immunity of adult dogs. PLoS One. 2020;15(8):e0238006. Published 2020 Aug 28. doi:10.1371/journal.pone.0238006

[138] Guarner F, Schaafsma GJ. Probiotics. Int J Food Microbiol. 1998;39(3):237-238. doi:10.1016/s0168-1605(97)00136-0

[139] Gallo A, Passaro G, Gasbarrini A, Landolfi R, Montalto M. Modulation of microbiota as treatment for intestinal inflammatory disorders: An uptodate. World J Gastroenterol. 2016;22(32):7186-7202. doi:10.3748/wjg.v22.i32.7186

[140] Hassan H, Rompola M, Glaser AW, Kinsey SE, Phillips RS. Systematic review and meta-analysis investigating the efficacy and safety of probiotics in people with cancer. Support Care Cancer. 2018;26(8):2503-2509. doi:10.1007/s00520-018-4216-z

[141] Rong J, Liu S, Hu C, Liu C. Single probiotic supplement suppresses colitis-associated colorectal tumorigenesis by modulating inflammatory development and microbial homeostasis. J Gastroenterol Hepatol. 2019;34(7):1182-1192. doi:10.1111/jgh.14516

[142] Maroof H, Hassan ZM, Mobarez AM, Mohamadabadi MA. Lactobacillus acidophilus could modulate the immune response against breast cancer in murine model. J Clin Immunol. 2012;32(6):1353-1359. doi:10.1007/s10875-012-9708-x

[143] Le Noci V, Guglielmetti S, Arioli S, et al. Modulation of Pulmonary Microbiota by Antibiotic or Probiotic Aerosol Therapy: A Strategy to Promote Immunosurveillance against Lung Metastases. Cell Rep. 2018;24(13):3528-3538. doi:10.1016/j.celrep.2018.08.090

[144] Naito S, Koga H, Yamaguchi A, et al. Prevention of recurrence with epirubicin and lactobacillus casei after transurethral resection of bladder cancer. J Urol. 2008;179(2):485-490. doi:10.1016/j.juro.2007.09.031

[145] Osterlund P, Ruotsalainen T, Korpela R, et al. Lactobacillus supplementation for diarrhoea related to chemotherapy of colorectal cancer: a randomised study. Br J Cancer. 2007;97(8):1028-1034. doi:10.1038/sj.bjc.6603990

[146] Motoori M, Yano M, Miyata H, et al. Randomized study of the effect of synbiotics during neoadjuvant chemotherapy on adverse events in esophageal cancer patients. Clin Nutr. 2017;36(1):93-99. doi:10.1016/j.clnu.2015.11.008

[147] Schmitz S, Suchodolski J. Understanding the canine intestinal

microbiota and its modification by pro-, pre- and synbiotics - what is the evidence?. Vet Med Sci. 2016;2(2):71-94. Published 2016 Jan 11. doi:10.1002/vms3.17

[148] White R, Atherly T, Guard B, et al. Randomized, controlled trial evaluating the effect of multi-strain probiotic on the mucosal microbiota in canine idiopathic inflammatory bowel disease. Gut Microbes. 2017;8(5):451-466. doi:10.1080/19490976.2017.1334754

[149] Rossi G, Pengo G, Caldin M, et al. Comparison of microbiological, histological, and immunomodulatory parameters in response to treatment with either combination therapy with prednisone and metronidazole or probiotic VSL#3 strains in dogs with idiopathic inflammatory bowel disease. PLoS One. 2014;9(4):e94699. Published 2014 Apr 10. doi:10.1371/journal.pone.0094699

[150] Xu H, Huang W, Hou Q, et al. Oral Administration of Compound Probiotics Improved Canine Feed Intake, Weight Gain, Immunity and Intestinal Microbiota. Front Immunol. 2019;10:666. Published 2019 Apr 2. doi:10.3389/fimmu.2019.00666

[151] Kao D, Roach B, Silva M, et al. Effect of Oral Capsule- vs Colonoscopy-Delivered Fecal Microbiota Transplantation on Recurrent *Clostridium difficile* Infection: A Randomized Clinical Trial. JAMA. 2017;318(20):1985-1993. doi:10.1001/jama.2017.17077

[152] Gough E, Shaikh H, Manges AR. Systematic review of intestinal microbiota transplantation (fecal bacteriotherapy) for recurrent *Clostridium difficile* infection. Clin Infect Dis. 2011;53(10):994-1002. doi:10.1093/cid/cir632

[153] Matson V, Fessler J, Bao R, et al. The commensal microbiome is associated with anti-PD-1 efficacy in metastatic melanoma patients. Science. 2018;359(6371):104-108. doi:10.1126/science.aao3290

[154] Gopalakrishnan V, Spencer CN, Nezi L, et al. Gut microbiome modulates response to anti-PD-1 immunotherapy in melanoma patients. Science. 2018;359(6371):97-103. doi:10.1126/science.aan4236

[155] Pereira GQ, Gomes LA, Santos IS, Alfieri AF, Weese JS, Costa MC. Fecal microbiota transplantation in puppies with canine parvovirus infection. J Vet Intern Med. 2018;32(2):707-711. doi:10.1111/jvim.15072

[156] Niederwerder MC. Fecal microbiota transplantation as a tool to treat and reduce susceptibility to disease in animals. Vet Immunol Immunopathol. 2018;206:65-72. doi:10.1016/j.vetimm.2018.11.002

[157] Koyasu S, Moro K. Role of innate lymphocytes in infection and inflammation. Front Immunol. 2012;3:101. Published 2012 May 7. doi:10.3389/fimmu.2012.00101

Chapter 7

Canine Detection of the Volatile Organic Compounds Related to Cervical Cancer Cells

Miriam Rodríguez-Esquivel, José Alejandro Sánchez-Pérez, Héctor Guerrero-Flores, Emmanuel Salcedo, Teresa Apresa-García, Ricardo López-Romero, Alejandra Valdivia-Flores, Florinda Jiménez-Vega, Rodrigo Aldariz and Mauricio Salcedo

Abstract

The use of trained dogs for the detection of volatile biomarkers in biological samples has great potential to be used for non-invasive diagnosis and monitoring of several diseases such as cancer. It offers early, highly accurate detection with fast response times, non-invasive to patients and allows for repeated sampling. The aforementioned methods are useful as a portable technology to increase detection, screening, and monitoring coverage in populations at risk. In this sense, Cervical Cancer (CC) has become a public health concern of alarming proportions in many developing countries, particularly in low-income sectors and marginalized regions due to different factors that limit the coverage of screening methods and the acceptance rates of women attending their routine gynecological examination. As such, early detection is a crucial medical factor in improving not only their population's quality of life but also its life expectancy. For the above, the great odor detection threshold exhibited by dogs is not unheard of and represents a potential opportunity to develop an affordable, accessible, and non-invasive method for detection of CC with high sensibility and specificity values.

Keywords: cervical cancer, dog detection, volatile organic compounds

1. Introduction

There is significant potential to reduce the suffering from cancer and to alleviate the economic burden to individuals, families, and societies. It is known that prevention campaigns and early detection interventions can avert cancer cases and deaths in high- and low-resource settings. Although many countries and communities have limited resources for screening, several common cancers among females such as Cervical Cancer (CC) have known means of prevention and/or early detection that can be applied in resource-appropriate settings [1].

Cervical cancer is one of the female reproductive system cancers, and it is a fundamental cause of cancer morbidity and mortality worldwide. This complex disease is relatively common with estimates of more than half a million new cases in 2018, and it accounts for 13% of all cancers in women in developed regions [2, 3]. The highest incidence rates (greater than 20 per 100,000 women) are found in Eastern, Western, and Southern Africa, South-Central Asia, South America, Melanesia, and Central Africa [3].

There are different tools to achieve CC elimination. The World Health Organization (WHO) has identified three critical targets to the elimination of this cancer type mainly in the increased coverage of: 1) Human Papillomavirus (HPV) vaccination, 2) Screening for premalignant disease with an HPV test, and appropriate management of women who screen positive, and 3) Reducing mortality from cervical cancer by providing appropriate treatment [4].

In recent years HPV vaccination in high-income countries has resulted in dramatic decreases in HPV infection and associated cervical disease as a primary prevention strategy for CC. Unfortunately, this has not happened in low- and middle-income countries where the access to the vaccination is limited mainly by the high cost, and therefore most women and girls at most risk cannot be protected. As a secondary prevention strategy, progress has been made in cervical precancer screening and treatment, but we must accelerate this momentum to reduce incidence and mortality worldwide to the meager rates found in wealthier countries [5]. In this sense, given that the access of the different CC prevention strategies is not equitable between countries or even inside of each country, due to the differences in infrastructure and access to health care systems so marked that we could find, it is necessary to search new tools and screening strategies. One of these strategies could be constituted by the markers present in the scent of CC cells.

The analysis of odors or volatile biomarkers emitted by cancer cells is of great value in the development of new diagnostic tests as low-risk methods for the early cancer diagnosis and a regular screening for all women, including the marginalized or disadvantaged. These volatile signatures are present in different biofluids and show a physiological status. In cancer, the analysis of these molecules has been demonstrated as a rapid and noninvasive alternative by analytical and biological ways as the use of trained dogs for the detection of several cancers.

Dogs can smell a trace of volatile odorous molecules or biomarkers (parts per trillions) emitted in different biofluids [6]. They have an extraordinary ability to recognizing odorous biochemical signature expressed only in ailing individuals but not in healthy individuals, in much earlier and better ways and with an accuracy comparable or superior to readily available sophisticated diagnostic instruments of the present time [6, 7]. The extraordinary canine sense of smell could avoid the unnecessary painful procedures on patients and minimize the time and expenditure on the diagnosis made through the biopsy and other tests having compromised sensitivity, specificity and predictive values resulting into inadequate accuracy. In CC, it could be an effective promissory weapon in fighting this disease and saving women's lives [7].

2. Cervical cancer as an important public health concern related to a virus infection

Among of the gynecological cancer types, CC must be the most detectable cancer due to access to the anatomical target. Unfortunately, this cancer type is the fourth most frequent cancer in the female population worldwide. More than 85% of cases occur in developing countries in which this malignancy is a public health concern due to its high mortality rates, provoked at least by late detection and lack of coverage for screening procedures [8]. Prior to the appearance of CC, women develop a precancer period that spans approximately two decades. During this extensive period, cervical epithelium cells present morphological and molecular changes that could not be typical of a healthy state neither of cancer; they are in transition state, as in the "limbo". Thus, these abnormal or precancer cells are known as squamous intraepithelial lesions (SIL) classified as low grade (LSIL) if they only affect the first third of the cervical epithelium, or high grade (HSIL) if they affect more than 50% of cervical epithelium layer [9].

Like other cancers, CC is a multifactorial disease, although there are different risk factors that could be controlled to prevent the development and progression of precursor lesions to invasive cancer. Among these risk factors are the onset of active sexual during the teen period, multiple sexual partners, overuse and uncontrolled of hormonal contraceptives, drug abuse, inadequate and overuse of antibiotic regimens, lack of protection during sexual intercourse and absence of routine gynecological inspections, among others; however, the main etiological factor associated with this type of cancer is the persistent HPV infection considered as sexually transmitted infection [10, 11].

The HPV is an infectious agent that is transmitted through sexual contact and affects the anus-genital and oropharyngeal tracts. This involves the transmission of one or more viral genotypes that infect the epithelia, which ones are classified as low risk (LR-HPV) and high risk (HR-HPV) according to their carcinogenic potential. Persistent infection by any of the HR-HPV genotypes is the cause of the vast majority of SIL [12]. It should be noted that the process of cellular transformation of normal cells to precancerous cells and invasive carcinoma involves a long period of time, as already mentioned, which makes the prevention of this neoplasm 100% feasible through the SIL screening.

3. CC and SIL screening

There are two approaches for CC screening and its precursor lesions. The first one is a cellular level approach (micro) involving cervical Pap smear cytology, and the second one is a tissue level approach (macro) through visual inspection iodine Lugol (VILI) or visual acid acetic inspection (VIA) and colposcopy [9].

A conventional Pap smear, which involves removing epithelial cells from the surface of the cervix with a brush (cytobrush or cervix brush) or spatula and then transferring them to a glass slide where they are prepared with Pap stain to be examined by a cytotechnologist or pathologist and discriminate between normal and abnormal cells by using conventional light microscopy. Colposcopy involves the inspection of the cervical tissue through a colposcope which allows magnification of the cervix up to 40X. The solutions already mentioned can be used to reveal changes in maturation, differentiation or abnormal epithelium vasculature that indicate the presence of SIL or CC [13]. It is worth mentioning that in the last years HPV testing is replacing cytology as the preferred cervical screening method; however, the interpretation of the HPV testing must be carried out with reservation [14].

HPV infection is one of the most prevalent sexually transmitted infections, generally symptomatic, with a worldwide prevalence in women with normal cytology of 11.4% and 99% in CC cases. Nevertheless, HPV infection is a necessary but not a sufficient cause of CC. Therefore, the positivity rate of HPV test is higher than cytology and that most positive test results do not indicate a high absolute risk of CC [15, 16]. These results could interpret as "false positive" CC screening results according to the use of this test as a screening tool for this cancer. This does not mean that HPV is not present; instead, we are referring to the detection of only an HPV infection that is not destined to cause CC [14].

4. Why is there CC?

As we know, CC is a preventable disease, and it has been shown that cervical cytology or Pap smear has been decreased the mortality rate of CC in developed countries. Unfortunately, this has not happened in low- and middle-income countries in which almost 9 out of 10 cervical cancer deaths occur, continuing as a priority health problem due to different social and technical factors involved [2].

CC screening program needs to be sufficiently accurate and acceptable for the target population by way of allowing the early detection of the disease and the triage of screen-positive women who requires colposcopy or treatment. To ensure the effectiveness of the screening, it is necessary a coverage rate of at least 80% of the population [2, 17]. Nevertheless, average Pap smear coverage is approximately 18.5% in developing countries, 63% in developed countries and 39.6% across the globe. These percentages could differ in each country but are clear that in any case, the coverage of Pap smear needs to be improved [18].

Pap smear and colposcopy are highly invasive methodologies since they require the introduction of a vaginal speculum to gain access to the cervix, thereby compromising the intimacy of the woman. Most of the screened women suffer shame, pain, inconvenience, or nervousness during the screening procedure, or can experience lower abdominal pain or vaginal bleeding in the days following the test. Women have reported a lack of information before or during gynecological inspections, and sometimes, they have referred a disrespectful attitude and a lack of engagement from the medical staff, resulting on women delaying their gynecological inspection or avoiding it altogether [19].

There are other social aspects interfering with the coverage of Pap screening in the risk population. Inadequate knowledge about the purpose and benefits of Pap smears, the fact that many screened and non-screened women do not know the meaning of an abnormal result. The fear and anxiety of having cervical abnormalities which affect the future decision to have a Pap test, social and health inequalities between women as a lack of health insurance, faults of organization in health-care programs involving in the appointment scheduling and the long waiting times to get a result, religious beliefs, taboo, fear of stigmatization, etc. [20, 21].

There are several technical limitations of the Pap smear screening as the specimen collection that imply collection and processing time, the procedure as the smear is taken, the quality of the samples, processing standards, lack of training of cytotechnologists for the accurate interpretation of results, the loss of concentration and fatigue that suffer by a repetitive task (up to 50 times a day) of visualization of slides. This provokes the rates of false-negative that can conduce to an increase in the cost CC screening and bad prognosis for the patients, as well as, the false-positive results that could cause psychological stress, overdiagnosis and overtreatment [9, 13].

5. OMIC era in the diagnosis of the diseases: volatilome and volatolome

To find alternative tools in cancer diagnosis in an earlier and more precise manner, researchers have explored the use of Metabolomics, specifically the volatile organic compounds (VOC) to detect these complex diseases.

VOC are carbon-based chemicals, volatile at room temperature and pressure, and source of most odors. Being produced during metabolic processes in millions of cells simultaneously, thus they are potentially releasing in an extracellular way on a detectable scale and may be emitted from different areas of the body prone to odor production e.g., scalp, axillae, feet, groin, oral cavity. These also can be excreted through different biofluids as saliva, breath, blood, sputum, feces, sweat, urine and may serve as ideal clinical biomarkers for several pathophysiological processes. The entire set of VOC produced by an organism is called Volatilome, and their accumulation inside and outside of the body reflects a unique metabolic state in an organism [6, 22, 23]. This knowledge is too old; the ancient Greek and Chinese human noses were the first to identify and describe the diagnostic potential of VOC in the diseases through the smell of different biological samples such as urine and sputum. Based on this ancestral knowledge, we know the VOC potential in the medical field, and ever since, our sense of smell has been used in medical practice as a more precise and less invasive diagnostic tool for the detection of several diseases [24]. Among the several diseases characterized by a specific odor are diabetes (rotten apple odor in the breath), scurvy (putrid body odor), cholera (rice water), trimethylaminuria (rotten fish-like odor in the breath, vaginal fluid, sweat and urine), phenylketonuria (musty odor), cystic fibrosis (chloride), or typhoid fever (baked bread body odor), etc. [25, 26].

Interestingly, in the last few decades, the diagnosis potential of VOC has focused on the search of the volatile profiles of many cancers in all the biofluids as the urine, feces, exhaled breath, and saliva of patients. The rationale for this is that cancer cells have different metabolic or biochemical requirements in comparison from normal cells, due to the genetic alterations that acquire and that allow them to proliferate outside the context of normal tissue development [27]. Therefore, the metabolic and bioenergetic alterations presented by tumor cells lead to a VOC profile different from that of healthy cells. These VOC profiles are useful for the diagnostic of cancer, predict patient response towards chemotherapies or treatment and monitor disease recurrences [28].

For example, the lack of sensitive and specific biomarkers for the early detection of prostate cancer led a Portuguese research group to investigate the performance of VOC present in the urine of patients as potential markers for this cancer in a metabolomic approach based on the analytical tool, the Gas Chromatography–Mass Spectrometry (GC–MS), finding a urinary profile of VOC different from that of cancer-free subjects with 78% sensitivity, 94% specificity and 86% accuracy [29].

A research group in the UK assessed the utility of VOC as feces biomarkers for colorectal neoplasia by headspace extraction followed by GC–MS. This group found that Propan-2-ol was the volatile organic compound most strongly associated with cancer, and 3-methylbutanoic acid or DL-menthol was the only volatile organic compound negatively associated with cancer. These VOC showed a diagnostic ability of sensitivity 87.9% and specificity 84.6% in the identification of colorectal adenocarcinoma [30].

Another example was research in which the gastric cancer was correlated with specific VOC biomarkers in the exhaled breath of a South American population. The exhaled VOC were analyzed by GC–MS and by a chemical gas sensor based on gold nanoparticles functionalized with octadecylamine ligands. Six VOC showed statistically significant differences between the cancer patients and the controls group (e.g., hexadecane and octadecane in the gastric cancer group, while eicosane and 1-cyclohexyl-2-(cyclohexylmethyl) pentane were identified as biomarkers in the control group). The sensor data responses to the breath samples yielded 97% accuracy, 100% sensitivity and 93% specificity [31].

Recently in Japan, the salivary metabolomic profile of oral squamous cell carcinoma was established through VOC analysis as potential biomarkers for the diagnosis of oral cancer through a method combining thin-film microextraction based on a ZSM 5/polydimethylsiloxane hybrid film coupled with GC–MS in saliva samples of oral cancer patients and healthy controls in which eighty kinds of volatile metabolites that were detected and identified and were classified as alcohols, ketones, hydrocarbons, aldehydes, organic acids, esters, phenols, etc. Among them, twelve COV were selected as potential oral cancer biomarkers for their use as non-invasive tools for the possible diagnosis of this cancer [32].

The above were just some examples of the vast literature that currently exists on the analysis of VOC for the diagnosis of different cancers through non-invasive samples and analytical methods.

6. Analytical chemistry (electronic noses) in VOC detection

As it was mentioned already, the analysis of VOC as cancer biomarkers in diverse biofluids is desirable because it allows a repeated sampling and a non-invasive and quick analysis. In this sense, there is great potential for the development and clinical application of VOC analysis in the diagnosis and monitoring of cancer [33].

The number of studies demonstrating the potential of VOC in cancer diagnosis has increased in the last decades due to analytical chemistry advancements that have made possible the quantitative analysis and comparison of VOC of cellular origin [6, 23].

From the several analytical techniques that exist, the GC–MS, Selected Ion Flow Tube Mass Spectrometry (SIFT-MS), Proton Transfer Reaction Mass Spectrometry (PTR-MS), Proton Transfer Reaction Time of Flight (PTR-ToF) and Ion Mobility Spectrometry (IMS) have been the most used to separate and identify VOC. These are sophisticated and valuables stationary analytical chemical instruments for the discovery of biological scents [34].

However, despite the low detection limits and high sensitivity offered by these methodologies, they present certain limitations that have prevented their routine application as screening methods. These require high levels of technical expertise and lengthy instrument run times (tens-of-minutes to hours) for detailed chemical analysis. Most of them, exceptionally high-resolution mass spectrometers, are extremely expensive and require expert maintenance. Furthermore, data interpretation, especially for non-targeted analyzes, may initially take many hours per sample until sufficient statistical results are accumulated to develop a targeted approach. Finally, they require infrastructure and trained personnel for their operation [35].

7. Dogs (biological noses) as clinical tools in cancer detection

Dogs have excellent odor detection capabilities in a vast range of fields. Their olfaction is a fundamental sense that let them perceive and comprehend the world around them. Humans have harnessed the canine sense of smell for the detection of different targets as an orthodox manner, such as explosives, land mines, narcotics, missing persons (forensic area), and invasive or endangered species [34, 36]. Right now, in this pandemic situation worldwide, dogs have been trained for COVID-19 early detection [37]. The question arises, why not use the canine olfactory for cancer detection?, nevertheless in the last decades, the use of canine olfaction as a diagnostic tool for identifying preclinical disease, especially cancer in biological samples has increased [34, 38].

Nowadays, there are a considerable number of publications using trained dogs to sort biological samples for follow-up and future diagnostics [31]. Several authors have published research suggesting that dogs can sort dozens of samples, including blind replicates and known control samples in a few minutes and may be able to detect lung, breast, prostate, ovarian, and melanoma cancers by smelling skin lesions, urine, exhaled breath, and surgically extracted tumors [35, 39].

The first report of dogs' potential to detect cancer was published in 1989 in the UK when a pet dog spontaneously detected its owner's melanoma, saving her life. After it, several additional cases of spontaneous cancer detection by dogs were reported, this caught the attention of the scientific community, and canine olfaction began to be used in the search for increasingly sensitive and specific diagnostic techniques for diseases as cancer where mass screening and early diagnosis could be improved [40].

In pilot work, a research group demonstrated the validity of using dogs as a biological system to examine exhaled breath in the diagnostic identification of lung and breast cancers. Its results showed an overall 99% sensitivity and specificity for canine scent detection among lung cancer patients and controls compared to biopsy-confirmed conventional diagnosis and 88% of sensitivity and 98% of specificity among breast cancer patients and controls [41].

Another research group in France, trained a Belgian Malinois shepherd by the clicker training method for prostate cancer detection on human urine samples. Its results showed that dogs could be trained to detect prostate cancer by smelling urine with a significant success rate (91% of sensitivity and specificity) suggesting that prostate cancer gives an odor signature to urine and it could be used as a potential screening tool [42].

In another research study, the ability of dogs to detect ovarian cancer from plasma samples was evaluated and how the odor associated to this cancer is affected by the treatment to reduce tumor burden, including surgery and five courses of chemotherapy. The dogs showed high sensitivity (97%) and specificity (99%) for the detection of ovarian cancer patients' plasma and indicated positive samples from patients who had recurrences. For the above, the dogs offer an outstanding assessment of ovarian cancer prognosis based on the specific odor in the blood which could enhance primary diagnosis and enable earlier relapse diagnosis and consequently an increase in patient's survival [43].

At present, Medical Detection Dogs in the UK an organization that is at the forefront of innovative research in the dogs' ability to detect the smell of human diseases and save lives. This organization focuses on detecting the VOCs associated with prostate cancer and colorectal cancer using trained dogs as a non-invasive method that can detect cancer at an early stage could both increase uptake of the screening and improve health outcomes [44].

Another current medical innovation research program is KDOG sustained by French Institute Curie (Paris) who is elaborating a simple, non-invasive, and cost-effective breast cancer screening method, based on canine detection. This method is contactless between the animal and the patient and the dogs' success rate in cancer detection that has been reported was about 100% [45].

8. Canine detection of cervical cancer

Issues concerning CC detection make it necessary searching for alternatives that help to increase the early screening coverage with greater percentages of sensitivity and specificity in screening and diagnostic tests. The introduction of methods capable of detecting virtually invisible -a single molecule among a billion or trillions of compounds- changes in the cell through the analysis of cells "odor" have much in their favor in practical applications. In this way, the key could rely upon the poorly explored field of the metabolomics of the cervicovaginal epithelium. Biotechnological and analytical systems such as dogs and analytics as GC–MS may be alternatives to current tests which leaves us with two good panoramas: 1) a laboratory analytical test; 2) a biotechnological field test; both of which use a "volatile biopsy", basically a scent sample, obtained without penetrating the human body at all. Our work team has proposed a device specifically made for this purpose [46, 47].

This device that our research group has developed is a gadget worn by the patient for a defined period, after which is simply stored in a container -provided by us- and mailed to the recipient, avoiding the stress of queuing in a hospital. Our device quickly collects in an *unorthodox manner* the VOC of the genitourinary tract allowing us, to use analytical devices GC–MS system for sample examination, with the surplus of being simple to dispose of after analysis is done. Each sample being scanned in one hour. Afterwards, the metabolic profile is reviewed by an expert to determine if there are any volatile biomarkers associated with the sample.

In this unorthodox scenario (for some people), this device is scanned by a trainer-dog binomial test carried out in seconds. In such an analysis, our gadget eliminates the shock some people could have by watching a dog "deciding their fate". Our results in both scenarios show that a sample's VOC profile result by the analytic test and detected by a trained dog, discriminates between cancerous and non-cancerous samples with more than a 90% sensitivity and specificity. These data are correlated afterwards to histopathological observation as the gold standard, suggesting that the device has a great value proposition [46, 47].

These proposals represent the ideal diagnostic tests for screening CC because they are non-invasive, low cost, accurate and partially portable, therefore meeting the requirements for a good screening test according to WHO. This established that a screening test must be sufficiently accurate to detect the condition earlier than in the absence of screening [48].

Recently another research group in Japan trained a dog to distinguish urine samples from cervical cancer patients from those of the controls, showing that cancer detection by dog sniffing can be a non-invasive, cost-effective screening technique for CC [49]. This report supports our proposal that the canine nose can be used and developed for CC detection.

The use of screening dogs is a real issue, for instance, they play vital roles helping in natural disasters or detecting drug or weapons trafficking as we have mentioned before; in the case of GC–MS itself is used again in the detection of drug trafficking, anti-doping or as a standard test in food products; then, why then should not they be used in the health-care industry? An example is the exhalomic test "Hearts Breath Test for Grade 3 Heart Transplant Rejection detection". Dr. Phillips et al. at Menssana Research Inc. in New Jersey USA developed this FDA approved test. This test detected a specific metabolomic profile and had opened a vast opportunity in the marketing of metabolomic or volatolomic tests [50].

9. A big challenge and an alternative

Pap test was not specifically developed to detect neither human cervical lesions nor HPV infection [44]. Moreover, it has not been subjected to a rigorous analysis regarding its sensitivity and specificity; however, it is the accepted test for detecting cytomorphological changes in the cervicovaginal epithelium but not for HPV. Epidemiological studies show that HPV detection does not necessarily indicate cancer, so it is considered necessary but not enough for CC development [51].

On the other hand, the CC research has served to define that prior to this type of invasive lesions, there are precursor or pre-invasive lesions as SIL, thanks to epidemiological studies, it has been determined that less than 10% of women infected with HPV will develop CC [52]. So far, it is unknown what factors are indispensable for the progression of these lesions causing *"headaches"* among oncologist and gynecologists alike, for nobody knows which SIL will progress to a more aggressive lesion. Many questions result in no concrete answer. Given these facts, the new OMICs area opens the opportunity for the early detection of pre-invasive lesions. Combining analytical and biotechnological procedures as aforementioned will permit the basis for new portable nanosystems.

In summary, our proposals are affordable and accessible for any women and could be an important weapon in this war against CC as a preventive measure deployed by health services, all this granting the public and government departments accept them.

10. Conclusions

Early detection saves lives. Therefore, it is necessary to implement new and alternative technologies that allow the development of accessible diagnostic methods that cover at least some of the limitations presented by conventional screening tests for the detection of diseases such as cancer, especially Cervical Cancer.

The canine detection of odors or volatile profiles emitted by cancer cells is a portable, highly sensitive and specific tool which could be used as a screening alternative (*as fast track*) in marginalized or areas of difficult access (even in the urban regions) to increase coverage in high-risk populations. Additionally, the use of a trained dog for screening could facilite prevention campaigns, saving money, time, labor, and lives due to an early diagnosis.

The CC and its precursor lesions detection is a priority health concern in different countries, therefore having an analytical alternative (GC–MS) and a biological alternative (canine smell) for screening could be a great support technique for conventional methods offering a non-invasive, fast, and accurate detection that can be carried out repeatedly and that would also be useful for monitoring the disease.

Applications for these tools extend to providing much needed medical attention for women from cultural backgrounds imposing several prohibitions, deep-rooted cultural taboos, religious beliefs, shame, or lack of health coverages. We thought that a "with a little help" to current methods by using improved, non-expensive and innovative procedures will conduct to accurate and timely diagnostics for this cancer type. Unfortunately, for both analytical and bio-detection methods, there is no consensus in the methodologies used or in the results obtained, thus this is a call to join forces with the scientific and social communities (research groups) for the replication of the studies that lead to the future implementation of these methodologies for clinical diagnosis.

Acknowledgements

The research team wants to thank every woman who accepted involvement in this project, an early Cervical Cancer diagnosis saves women. To oncologists, gynecologists, and students who collaborated with us to 'put a grain of sand'. In the last moment Miriam Rodriguez was considered as co-corresponding. Finally, to IMSS-CIS and Dr. Tomás Hernández-Quijano for support our research and new approach for our women.

Author details

Miriam Rodríguez-Esquivel[1*], José Alejandro Sánchez-Pérez[2], Héctor Guerrero-Flores[3], Emmanuel Salcedo[1], Teresa Apresa-García[1], Ricardo López-Romero[1], Alejandra Valdivia-Flores[4], Florinda Jiménez-Vega[5], Rodrigo Aldariz[1] and Mauricio Salcedo[1*]

1 Medical Research Unit in Oncological Diseases, Oncology Hospital, National Medical Center Century XXI, Mexican Institute of Social Security, Mexico City, Mexico

2 Canine Detector Group Mexico, Mexico City, México

3 Mexican Institute of Social Security, Southern Delegation Medical Coordinator, Mexico City, Mexico

4 Instituto Politécnico Nacional, Escuela Superior de Enfermería y Obstetricia, Mexico City, México

5 Institute of Biomedical Sciences, Autonomous University of Juarez City, Juarez City, Mexico

*Address all correspondence to: miririamre@gmail.com and masava89@gmail.com

References

[1] Torre LA, Islami F, Siegel RL, Ward EM, Jemal A. Global Cancer in Women: Burden and Trends. Cancer Epidemiol Biomarkers Prev. 2017;26(4): 444-58.

[2] Porcari AM, Negrão F, Tripodi GL, Pitta DR, Campos EA, Montis DM, et al. Molecular Signatures of High-Grade Cervical Lesions. Front Oncol. 2018;8(99):1-9.

[3] Weiderpass E, Labrèche F. Malignant Tumors of the Female Reproductive System. Saf Heal Work. 2012;3:166-80.

[4] Adanu RMK. Eradicating cervical cancer. Int J Gynecol Obs. 2020;148:1-2.

[5] Aranda S, Berkley S, Cowal S, Dybul M, Evans T, Iversen K, et al. Ending cervical cancer: A call to action. Int J Gynecol Obs 2017; 2017;138:4-6.

[6] Angle C, Waggoner LP, Ferrando A, Haney P, Otto CM, Goldblatt A, et al. Canine detection of the Volatilome: a review of implications for pathogen and disease detection. Front Vet Sci. 2016;3(47):1-7.

[7] Ranjan K, Singh R. Dog Nose to E-Nose in Disease Diagnosis Koushlesh. J Adv Biol. 2018;11:2294-306.

[8] LaVigne AW, Triedman SA, Randall TC, Trimble EL, Viswanathan AN. Cervical cancer in low and middle income countries: Addressing barriers to radiotherapy delivery. Gynecol Oncol Reports [Internet]. 2017;22:16-20. Available from: https://doi.org/10.1016/j.gore.2017.08.004

[9] Jusman Y, Ng SC, Osman NAA. Intelligent Screening Systems for Cervical Cancer. Sci World Journa. 2014;2014:1-15.

[10] Gakidou E, Nordhagen S, Obermeyer Z. Coverage of cervical cancer screening in 57 countries: Low average levels and large inequalities. PLoS Med. 2008;5(6):0863-8.

[11] Petry KU, Wörmann B, Schneider A. Benefits and risks of cervical cancer screening. Oncol Res Treat. 2014;37:48-57.

[12] Shah SS, Senapati S, Klacsmann F, Miller DL, Johnson JJ, Chang HC, et al. Current technologies and recent developments for screening of HPV-associated cervical and oropharyngeal cancers. Cancers (Basel). 2016;8(9):1-27.

[13] Coleman D. Limitations of the cervical smear test as a method of detecting women at risk of cervical cancer. AVMA Med Leg J Limitations. 2015;7(6):235-40.

[14] Schiffman M, Sanjose S de. False positive cervical HPV screening test results. Papillomavirus Res J. 2019;7:184-7.

[15] Coscia MF, Monno R, Ballini A, Mirgaldi R, Dipalma G, Pettini F, et al. Human papilloma virus (HPV) genotypes prevalence in a region of South Italy (Apulia). Ann Ist Super Sanità. 2015;51(3):248-51.

[16] Kops NL, Bessel M, Horvath JDC, Domingues C, Souza FMA de, Benzaken AS, et al. Factors associated with HPV and other self-reported STI coinfections among sexually active Brazilian young adults: cross-sectional nationwide study. BMJ Open. 2019;1-7.

[17] Sousa DMDN, Chagas ACMA, Vasconcelos CTM, Stein AT, Oriá MOB. Development of a clinical protocol for detection of cervical cancer precursor lesions. Rev Latino-Am Enferm. 2018;26(e2999).

[18] Ranabhat S, Dhungana G, Neupane M, Shrestha R, Tiwari M. Pap

smear coverage and effect of knowledge and attitude regarding cervical cancer on utilization of the test by women in Udayapur District of Nepal. J Chitwan Med Coll. 2014;4(10):31-5.

[19] Korfage IJ, Ballegooijen M Van, Wauben B, Looman CWN, Habbema JDF, Essink-bot M. Having a Pap smear , quality of life before and after cervical screening: a questionnaire study. BJOG. 2012;119(8):936-44.

[20] Paolino M, Arrossi S. Women ' s Knowledge About Cervical Cancer , Pap Smear and Human Papillomavirus and Its Relation to Screening in Argentina. Women Health. 2011;51:72-87.

[21] Rees HD, Lombardo AR, Tangoren CG, Meyers SJ, Muppala VR, Niccolai LM. Knowledge and beliefs regarding cervical cancer screening and HPV vaccination among urban and rural women in León , Nicaragua. PeerJ. 2017;1-34.

[22] Whittle CL, Fakharzadeh S, Eades J, Preti G. Human Breath Odors and Their Use in Diagnosis. Ann N Y Acad Sci. 2007;266:252-66.

[23] Berkhout DJC, Benninga MA, Boer NKH De, Meij TG De. Revival of an ancient Greek art : scent detection as diagnostic tool for tuberculosis. Pediatr Res. 2018;84:4-5.

[24] Pomerantz A, Blachman-braun R, Galnares-olalde JA, Berebichez-fridman R, Capurso-garcía M. The possibility of inventing new technologies in the detection of cancer by applying elements of the canine olfactory apparatus. Med Hypotheses. 2015;85(2):160-72.

[25] Shirasu M, Touhara K. The scent of disease: volatile organic compounds of the human body related to disease and disorder. J Biochem. 2011;150(3):257-66.

[26] Bijland LR, Bomers MK, Smulders YM. Smelling the diagnosis A review on the use of scent in diagnosing disease. Neth J Med. 2013;71(6):300-7.

[27] Locasale JW, Cantley LC. Altered metabolism in cancer. BMC Biol. 2010;8(88):1-3.

[28] Lubes G, Goodarzi M. GC-MS based metabolomics used for the identification of cancer volatile organic compounds as biomarkers. J Pharm Biomed Anal. 2018;147:313-22.

[29] Lima AR, Pinto J, Azevedo AI, Barros-silva D, Jerónimo C, Henrique R, et al. Identification of a biomarker panel for improvement of prostate cancer diagnosis by volatile metabolic profiling of urine. Br J Cancer. 2019;121(10):857-68.

[30] Bond A, Greenwood R, Lewis S, Corfe B, Sarkar S, O'Toole P, et al. Volatile organic compounds emitted from faeces as a biomarker for colorectal cancer. Aliment Pharmacol Ther. 2019;49:1005-12.

[31] Durán-acevedo CM, Jaimes-mogollón AL, Oscar, Gualdrón-Guerrero E, Welearegay TG, Martinez-Marín JD, et al. Exhaled breath analysis for gastric cancer diagnosis in Colombian patients. Oncotarget. 2018;9(48):28805-17.

[32] Shigeyama H, Wang T, Ichinose M, Ansai T, Lee S. Identification of volatile metabolites in human saliva from patients with oral squamous cell carcinoma via zeolite-based thin-film microextraction coupled with GC – MS. J Chromatogr B. 2019;1104:49-58.

[33] Dummer J, Storer M, Swanney M, Mcewan M, Scott-thomas A, Bhandari S, et al. Analysis of biogenic volatile organic compounds in human health and disease. Trends Anal Chem. 2011;30(7):960-7.

[34] Hackner K, Pleil J. Canine olfaction as an alternative to analytical

instruments for disease diagnosis: understanding "dog personality" to achieve reproducible results. J Breath Res. 2018;11(1):1-8.

[35] Pleil JD, Wallace MAG, Mccord J, Madden MC, Sobus J, Ferguson G. How do cancer-sniffing dogs sort biological samples ? Exploring case-control samples with immunochemistry methods. J Breath Res. 2020;14(1):1-17.

[36] Fischer-tenhagen C, Johnen D, Heuwieser W, Becker R, Schallschmidt K, Nehls I. Odor Perception by Dogs: Evaluating Two Training Approaches for Odor Learning of Sniffer Dogs. Chem Senses. 2017;42:435-41.

[37] Grandjean D, Sarkis R, Lecoq-Julien C, Aymeric Benard, Roger V, Levesque E, et al. Can the detection dog alert on COVID-19 positive persons by sniffing axillary sweat samples? A proof-of-concept study. PLoS One. 2020;15(12):1-19.

[38] Moser AY, Bizo L, Brown WY. Olfactory Generalization in Detector Dogs. Animals. 2019;9(702):1-12.

[39] Moser E, Mcculloch M. Canine scent detection of human cancers : A review of methods and accuracy. J Vet Behav Clin Appl Res. 2010;5(3):145-52.

[40] Elliker K, Williams H. Detection of skin cancer odours using dogs: a step forward in melanoma detection training and research methodologies. Br J Dermatol. 2016;175(5):851-2.

[41] Mcculloch M, Jezierski T, Broffman M, Hubbard A, Turner K, Janecki T. Diagnostic Accuracy of Canine Scent Detection in Early- and Late-Stage Lung and Breast Cancers. Integr Cancer Ther. 2006;5(1):30-9.

[42] Cornu J-N, Cancel-Tassin G, Ondet V, Girardet C, Cussenot O. Olfactory Detection of Prostate Cancer by Dogs Sniffing Urine: A Step Forward in Early Diagnosis. Eur Urol. 2011;59:197-201.

[43] Horvath G, Andersson H, Nemes S. Cancer odor in the blood of ovarian cancer patients: a retrospective study of detection by dogs during treatment, 3 and 6 months afterward. BMC Cancer. 2013;13(396):1′7.

[44] Cancer detection - Medical Detection Dogs [Internet]. [cited 2020 Dec 18]. Available from: https://www.medicaldetectiondogs.org.uk/cancer-detection/

[45] Institut Curie | KDOG Cancer Detect Group [Internet]. [cited 2020 Dec 18]. Available from: https://kdog.institut-curie.org/

[46] Guerrero-Flores H, Apresa-García T, Garay-Villar Ó, Sánchez-Pérez A, Flores-Villegas D, Bandera-Calderón A, et al. A non-invasive tool for detecting cervical cancer odor by trained scent dogs. BMC Cancer. 2017;17(79):1-8.

[47] Rodríguez-Esquivel M, Rosales J, Castro R, Apresa-García T, Garay O, Romero-Morelos P, et al. Volatolome of the Female Genitourinary Area: Toward the Metabolome of Cervical Cancer. Arch Med Res. 2018;49(1):27-35.

[48] Arbyn M, Weiderpass E, Bruni L, Sanjosé S De, Saraiya M, Ferlay J, et al. Estimates of incidence and mortality of cervical cancer in 2018: a worldwide analysis. Lancet. 2020;8:191-203.

[49] Yamamoto A, Kamoi S, Kurose K, Ito M, Takeshita T, Kure S, et al. The trained sniffer dog could accurately detect the urine samples from the patients with cervical cancer, and even cervical intraepithelial neoplasia grade 3: A pilot study. Cancers (Basel). 2020;12(11):1-12.

[50] Phillips M, Boehmer JP, Cataneo RN, Cheema T, Greenberg J,

Kobashigawa J, et al. Heart Allograft Rejection: Detection With Breath Alkanes in Low Levels (the HARDBALL Study). J Hear Lung Transplant. 2004;23(6):701-8.

[51] Walboomers JMM, Jacobs M V., Manos MM, Bosch FX, Kummer JA, Shah K V., et al. Human Papillomavirus is a necessary cause of invasive cervical cancer worldwide. J Pathol. 1999;189:12-9.

[52] Castellsagué X. Natural history and epidemiology of HPV infection and cervical cancer. Gynecol Oncol. 2008;110:S4-7.

Chapter 8

Canine Hepatic Carcinoma: Diagnoses and Treatments Via Global State-of-the-Art Approach and Traditional Chinese Veterinary Medicine

Chih-Feng Chen, Shao-Wen Chiu, Lung-Kwang Pan and Chao-Hsun Chuang

Abstract

This chapter discusses effective diagnostics and treatment of canine hepatic carcinoma (CHC), where state-of-the-art global technologies are complemented by traditional Chinese veterinary medicine (TCVM). The biokinetic Ga-67 model of CHC is proposed to clarify the Ga-67 metabolic mechanism among various organs. It is aimed at identifying the best routine for detecting the metastatic or primary CHC and substantiating the optimal further treatment. The routine examination of CHC can be performed via Ga-67 nuclear examination or MRI, biological index, X-ray, and abdominal ultrasound. The available methods of animal cancer treatment imply separate or combined application of surgery, radiation therapy, and chemotherapy targeted at the particular cancer cells. However, there is also a general concern on the quality of life of pets/canine patients. This leaves enough space to the TCVM (including acupuncture and famous herbal drugs) with a long application history in Asia and growing usage as alternative treatment in other regions. However, its current applications to domestic animals/pets suffering from carcinomas are based on individual expert opinions, while there are no outlined veterinary treatment strategies and guidelines for clinical practice in this field. A comprehensive combination of state-of-the-art global technologies and TCVM is considered instrumental in curing canine hepatic carcinoma.

Keywords: canine hepatic carcinoma, nuclear examination, Ga-67 biokinetic model, traditional Chinese veterinary medicine, acupuncture

1. Introduction

1.1 Motivation and purpose

Dog owners are aware of the sad fact that their pets are vulnerable to various tumors that may occur at any age/location and cause severe complications. In particular, canine hepatic tumors correspond to 0.6 to 1.3% of canine neoplasms, while about 7–36% of dogs are metastasized from other organs [1–9]. The canine patients, which usually suffer from systemic metabolic stress and cachexia, require a surgery or chemotherapy based on the veterinarians' recommendations and life expectations. Noteworthy is that different tumor types of primary carcinoma can arise from different cell types listed in **Table 1** [2]. The major types of primary liver cancer are hepatocellular carcinoma (HCC), bile duct carcinoma, neuroendocrine (carcinoid) tumor, and mesenchymal tumor (sarcoma). Hepatocellular carcinoma (HCC) is generally found in dogs, especially elder ones. Compared with the modular or diffuse forms, the majority of HCC has a lower rate of metastasis. A single massive HCC tumor can usually be removed by surgical treatment. However, the most difficult clinical treatment is the diffuse HCC involving in the entire liver. Dogs with multiple liver lobes' HCC are not recommended to undergo surgical resection, have a poor prognosis and very limited treatment options [2–4, 10–13]. Bile duct carcinomas are the second common malignant liver tumor in dogs. Neuroendocrine tumors are quite rare and mostly nodular or diffuse, while primary liver sarcomas, including hemangiosarcoma, fibrosarcoma, and hepatocellular carcinoma, are unusual clinically. Dogs with liver tumors can be either asymptomatic or exhibit nausea, vomiting, weight loss, loss of appetite, diarrhea, lethargy, or PU/PD. Occasionally, yellowing of the skin and eyes like jaundice, or neurological signs (hepatic encephalopathy), such as seizures, disorientation, and weakness, are observed. Liver carcinoma grows slowly and manifests itself too late [2–6, 14–16]. This study attempts to set up the Ga-67 nuclear examination protocol for liver carcinoma canine patients, which would guide the veterinary expert and dog owner on the further optimal treatment: intensive surgery/chemotherapy or/and traditional Chinese veterinary medicine.

1.2 Definition of critical terminology

In this study, the global state-of-the-art approach to the above problem implies the conventional medicine based on modern science and advanced evaluation methods of canine physical and biochemical conditions. The respective treatment methods include chemotherapy, drugs, radiology, and surgery. Local tumor ablation and radiotherapy have been applied in veterinary medicine for decades. Chemotherapy is related to the use of several anti-cancer drugs administered intravenously, orally, or subcutaneously, which circulate in the patient's body and attack cancer cells. The most common chemotherapy drugs for treating liver cancer are listed in **Table 2**. Alkylating agents react with DNA strands and change the DNA structure. The commonly utilized

Classification	Types
1. Hepatocellular	a. Hepatic adenoma b. Hepatocellular carcinoma c. Hepatoblastoma
2. Biliary	a. Biliary adenoma (or cystadenoma) b. Biliary carcinoma
3. Neuroendocrine	a. Neuroendocrine carcinoma or carcinoid
4. Mesenchymal	a. Hemangiosarcoma b. Leiomyosarcoma c. Fibrosarcoma d. Osteosarcoma e. Malignant mesenchymoma f. Chondrosarcoma

Source: DOI: 10.1016/j.cvsm.2016.11.016.

Table 1.
Primary hepatobiliary tumor types.

Drug	**Indications**	**Toxicity**	**Dosage and Route of Administration**
Alkylating Agents			
Cyclophosphamide	Lymphoma, Sarcoma, Carcinoma	Sterile hemorrhagic cystitis BM[*], GI[#]	Typically 200–250 mg/m^2 IV or PO
Chlorambucil	Lymphoma (lymphocytic) CLL[§], MCT[※], to replace Cyclophosphamide if sterile hemorrhagic cystitis occurs	BM[*] (mild)	Variable PO
CCNU (Lomustine)	Lymphoma, MCT[※]	BM[*], liver	Dogs: 60–90 mg/m^2 PO q3wk Cats: 50 to 60 mg/m^2 PO or 10 mg/cat PO q3 to 4wk
Melphalan	MM[Ψ], Plasma cell tumors	BM[*]	0.1 mg/kg/day PO x 10 days, then 0.05 mg/kg QOD; 0.2 pulse dose at 7 mg/m^2 daily for five days q3wk
Anthracyclines			
Doxorubicin	Lymphoma, Sarcoma, Carcinoma	BM[*], GI[#], hypersensitivity reaction, perivascular damage with extravasation, cumulative myocardial toxicity (dogs), nephrotoxicity (cats)	Dogs:≥15 kg: 30 mg/m^2 IV q2 to 3wk Dogs: < 15 kg: 1 mg/kg IV q2 to 3wk Cats: 1 mg/kg or 25 mg/m^2 q3wk
Mitoxantrone	Lymphoma Carcinoma	BM[*], GI[#], perivascular damage with extravasation	Dogs: 5 to 6 mg/m^2 IV q3wk Cats: 6 to 6.5 mg/m^2 IV q3wk
Platinum Drugs			
Carboplatin	Sarcoma, Carcinoma	BM[*], GI[#]	Dogs: 300 mg/m^2 IV q3wk Cats: 240 to 260 mg/m^2 IV q3 to 4wk
Cisplatin	Sarcoma, Carcinoma	BM[*], Nephrotoxicity, Fatal pulmonary edema (cats)	Dogs: 70 mg/m^2 IV q3wk (saline Do not use in cats emesis, diuresis)
Antimetabolites			
Methotrexate	Lymphoma	BM[*] (mild), GI[#]	0.8 mg/kg IV once weekly
Cytosine arabinoside	Lymphoma	BM[*] (mild), GI[#]	Variable; SQ, IM. IV.
Gemcitabine	Lymphoma, MM[Ψ], Carcinoma. feline SCC[¶]	BM[*]	Limited studies Dogs: 100 to 1000 mg/m^2 IV; 50 mg/m^2 twice/wk. with RT[ω] Cats: 25 mg/m^2 twice/wk. with RT[ω]

Drug	Indications	Toxicity	Dosage and Route of Administration
5FU	Sarcoma, carcinoma	BM*, GI#, Neurotoxicity, Fatal neurotoxicity in cats	Dogs: 150 mg/m^2 once/ wk. IV, topically Do not use in cats
Antimicrotubule Agents			
Vincristine	Lymphoproliferative cancer, Sarcoma	BM*, GI#, Constipation (ileus), Peripheral neuropathy, Perivascular tissue reaction	0.5 to 0.75 mg/m^2 IV weekly or every other week
Vinblastine	Lymphoma, MCT⋇	BM*, GI#, Perivascular tissue reaction	2 to 2.2 mg/m^2 IV weekly
Miscellaneous			
L-Asparaginase	Lymphoma	Hypersensitivity reaction	400 IU/kg or 10.000 IU/ m^2 SQ or IM (10,000 IU maximum) weekly
Procarbazine	Lymphoma	BM*, GI#	Variable; dogs: 50 mg/ m^2/day

**: BM; bone marrow, #: Gl; gastrointestinal, ʆ: CLL; chronic lymphocytic leukemia, ⋇: MCT; mast cell tumor, Ψ: MM; multiple myeloma, ω: RT: radiation therapy, ¶: SCC; squamous cell carcinoma.*
Source: https://veteriankey.com/chemotherapy/, chapter 44: Chemotherapy.
Alkylating agents react with DNA strands and changing the DNA structure. Anthracycline antibiotics exert their cytotoxic effect through different mechanisms, including free radical formation, DNA intercalation, and protein synthesis inhibition. Platinum drugs act at the N7 position of guanine and adenine residues, suggesting DNA as the primary target of the drug. Antimetabolites are drugs that interfere with enzymes or affect DNA synthesis. Antimicrotubule Agents interfere with the mitotic spindle, which inhibits cellular division and proliferation.

Table 2.
Classification of chemotherapy drugs for treatment.

alkylating agents in veterinary oncology are chlorambucil, cyclophosphamide, melphalan, and lomustine. The common toxicity of cyclophosphamide is related to sterile hemorrhagic cystitis. Anthracycline antibiotics exert their cytotoxic effect through different mechanisms, including free radical formation, DNA intercalation, and protein synthesis inhibition. An example of such medicine is Mitoxantrone, which is metabolized in the liver and eliminated via feces and urine. Platinum drugs act at the N7 position of guanine and adenine residues, having DNA as their primary target and being eliminated via kidneys [6]. Antimetabolites are drugs that interfere with enzymes or affect DNA synthesis. Their chemo group, methotrexate, is primarily eliminated via renal excretion. Finally, there are antimicrotubule agents, which interfere with the mitotic division process, inhibiting cell division and proliferation.

Quite a recent treatment introduced to veterinary oncology is radiology or radiation therapy. It rapidly became one of the most common procedures of cancer treatment, insofar as it could be clinically performed when surgical resection of the tumor remaining after chemotherapy would be problematic or too dangerous. Radiotherapy can make cells unable to replicate and eliminate the cell division and proliferation. However, radiation therapy can damage normal cells, rather than only malignant ones. A well-designed radiotherapy program is intended and designed to maximize tumor effect and minimize normal tissue effect. The benefit of radiotherapy is a cure for tumors and a split course treatment before surgery or chemotherapy [17].

The alternative approach to the above "targeted instrumental and drug invasion into the injured body part" is the overall improvement of the whole body and emotional well-being of a patient. In case of canine patients, this approach is reduced to the Traditional Chinese Veterinary Medicine (TCVM), which has been introduced several thousand years ago as a system of health care and is being implemented globally nowadays as alternative medicine [12]. Briefly, the TCVM has accumulated a vast list of robust diagnoses of canine diseases based on basic symptoms and appearances (tongue color, skin conditions, iris, pulse, etc.). A veterinarian analyzes these symptoms and the related overall system problems. The proposed cure is TCVM-based herbal medicine aimed to fine-tune the immune system and prevent a reoccurrence of an illness or future problem as illustrated in **Figure 1** [18, 19].

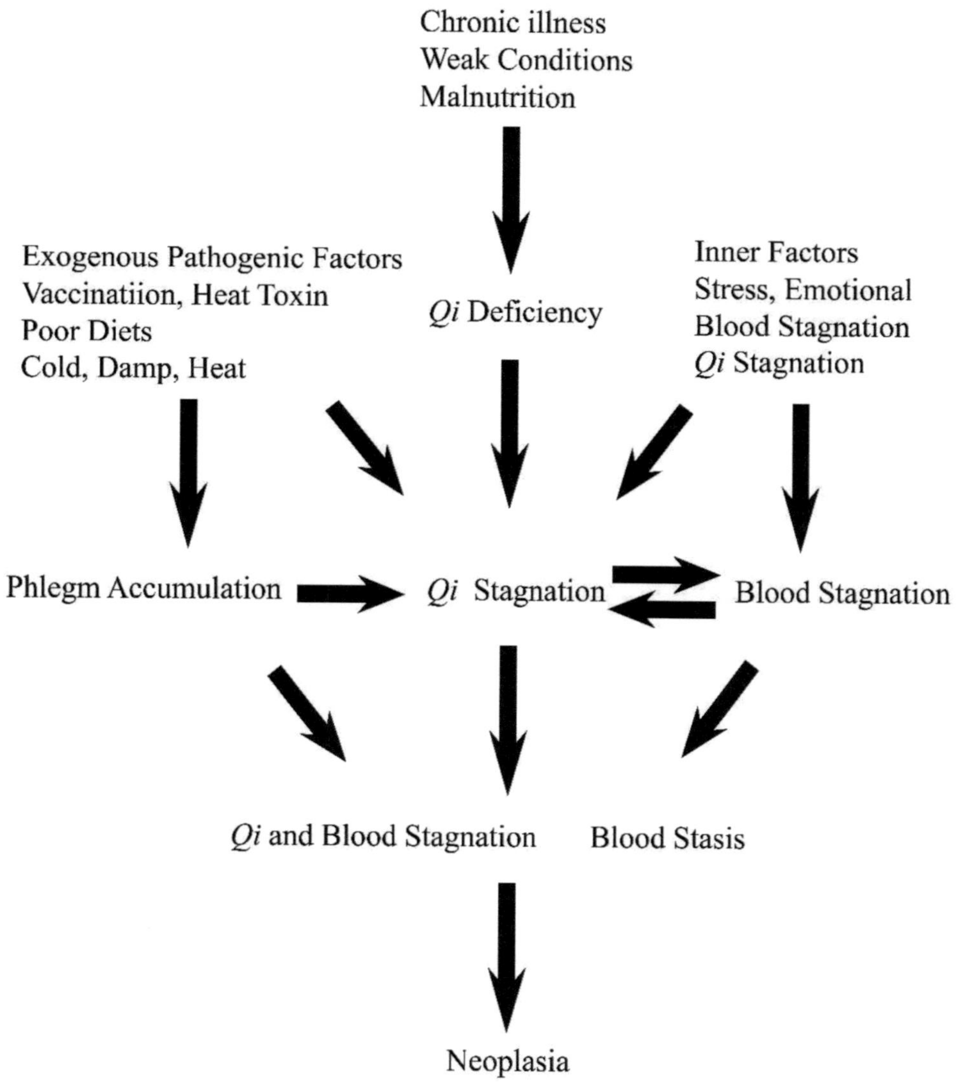

Figure 1.
Etiology of neoplasia in TCVM. The factors contributing to form phlegm, qi and blood stagnation, blood stasis, and eventually, neoplasia, are summarized. It is shown that qi deficiency is the root of cancer and phlegm, qi/blood stagnation, and blood stasis are one part of the branches. (cited from Huisheng Xie, Vanessa Preast, TCVM fundamental principles, 2nd edition).

1.3 Background review and rationale study with reference

Liver carcinoma can be diagnosed or detected by using multiple methods. Histological findings are considered the most robust for canine patients. The TNM staging system for canine hepatocellular carcinoma (HCC) is summarized in **Table 3** [20]. Veterinary surgeons may perform additional blood examinations to look for signs of liver dysfunction. Surgery is one of the treatments for all liver tumors. However, right-sided tumors are more challenging to resect because the resection lesion is related to the vena cava [2]. For massive hepatic carcinoma, surgical resection via lobectomy is the treatment when complete resection is accepted [10, 11]. The mass ligation for complete lobectomy is not recommended for large dogs, or central or right divisional liver tumors, because this method will increase the risk of complications, such as bleeding or bile leakage. Surgical stapling devices are recommended to perform liver lobectomy; in these devices, overlapping rows of staples are quickly placed to attenuate vascular and biliary structures within the liver lobe's hilus. If stages T3, N1, or M1 in **Table 3** are confirmed by clinical-stage evaluations, no surgical resection is recommended [20] because it would not remove some malignant liver tumors. Therefore, chemotherapy becomes an alternative treatment. Unfortunately, according to many clinical practices and references, chemotherapy is not very effective liver carcinoma treatment. Thus, HCC treatment with intravenous gemcitabine or carboplatin is no longer performed. Instead, in the last decade, the Metronomic chemotherapy and Sorafenib treatment (5 mg/kg, twice daily) became widely used due to lower-dose usage and administration ease [6, 7].

Alternatively, radiotherapy is sometimes used to make the liver tumor smaller or incapsulate it. However, it is not applicable to most liver tumor cases because of its side-effects. The major complication is a radiation heat-induced damage to the adjacent unaffected liver tissue. A 3D-CRT (three-dimensional conformal radiation therapy) is introduced as a new viable treatment option for canine patients with an inoperable massive liver carcinoma. From 6 to 10 Gy per fraction are prescribed on the planning target volume, and the total dose is 18–42 Gy with 1 to 2 fractions per week [17].

1.4 The innovative features of this study

This study is the first attempt to apply the biokinetic Ga-67 model to canine liver carcinoma. The aim is to identify the best routine for detecting the metastatic or primary hepatic carcinoma and substantiating the optimal further treatment.

Classification	Stage
Primary tumor (T)	T0: no evidence of primary tumor Tl: solitary tumor of any size involving one lobe T2: multiple tumors of any size involving multiple lobes T3: tumor(s) with direct invasion of adjacent organs regional lymph nodes
Regional lymph nodes (N)	N0: no regional lymph nodes metastasis Nl: regional lymph node metastasis N2: lymph node metastasis Distant metastasis
Distant metastasis (M)	M0: no distant metastasis Ml: distant metastasis

Source: doi:10.3390/cancers12051272

Table 3.
The TNM staging system for canine hepatocellular carcinoma.

1.5 The specific rationale study with a solid description

The biokinetic model of Ga-67 evolution was elaborated in this study for the case–control group of canine liver carcinoma via in-vivo gamma camera/8-slice CT technique. The circulation of time-dependent concentrations of Ga-67 among organs was monitored and simulated. The obtained quantitative data for organs and branching ratios among organs were incorporated into the biokinetic model of Ga-67 radionuclide administered during the hepatic survey.

2. Diagnosis of canine hepatic carcinoma

Due to the absence of nerves in the liver, the early liver neoplasia is painless. Therefore, when canine patients are clinically examined, their liver disease if any is diagnosed as moderate or severe.

The clinical symptoms include depression, lack of appetite, vomiting, weight loss, diarrhea, PU/PD, abdominal distention, lethargy, icterus, and ascites. The neurological disorder is mainly caused by hypoglycemia, hepatic encephalopathy, or metastasis of the central nervous system as shown in **Figure 2** [9, 21–24].

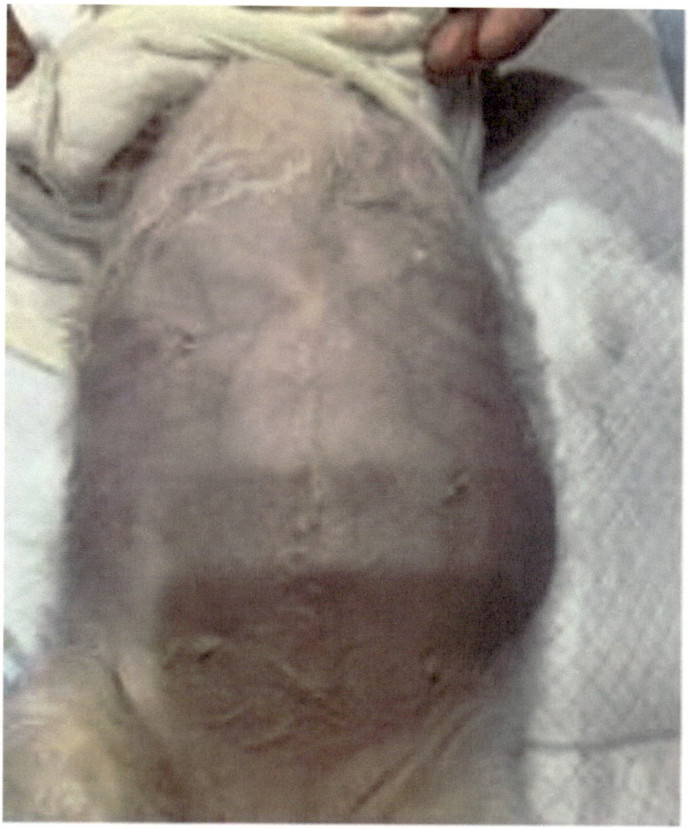

Figure 2.
A 12-year-old female Maltese had distention of the abdomen, was diagnosed with a liver tumor via ultrasound.

2.1 Biological index

Hematologic features, including mild non-regenerative anemia, inflamed leukocytosis, and thrombocytosis, are common in canine liver tumors [1, 15, 16].

Thrombocytosis is observed in approximately half of canine patients with massive HCC [1]. The prolonged blood coagulation (prothrombin time, thrombin time, and activated partial thromboplastin time) is rare unless the liver disfunction occurs, being observed in only 20% of HCC cases [1, 16].

Serum biochemistry indexes of dogs with hepatic neoplasia are commonly increased. Thus, elevated enzyme alanine transaminase (ALT), alkaline phosphatase (ALP), and aspartate aminotransferase (AST) are observed in liver tumors. Moreover, AST to ALT ratios below one are consistent with HCC or bile duct carcinoma [1].

The increased content of α-Fetoprotein is used as a robust indicator of HCC in human patients [23, 24]. This indicator is less suitable for dogs with hepatic neoplasia, being intrinsic to other types of canine liver tumors. However, about 75% of dogs with HCC had increased serum α-Fetoprotein, as reported in [25, 26]. Hypoalbuminemia, hypoglycemia, hyperglobulinemia, and elevated bile acid concentration are also observed in canine liver tumors.

2.2 X-ray examination

Regular abdominal radiography images of canine patients obtained via a diagnostic X-ray device are used for diagnosing, staging liver tumors, and planning the surgery operation if any. Abdominal radiography may identify a cranial abdominal mass with caudal and lateral recumbency shift of the stomach. Ascites may interfere with visualizing a mass as depicted in **Figure 3**. Assessment of thoracic radiographs is considered critical to exclude metastatic disease.

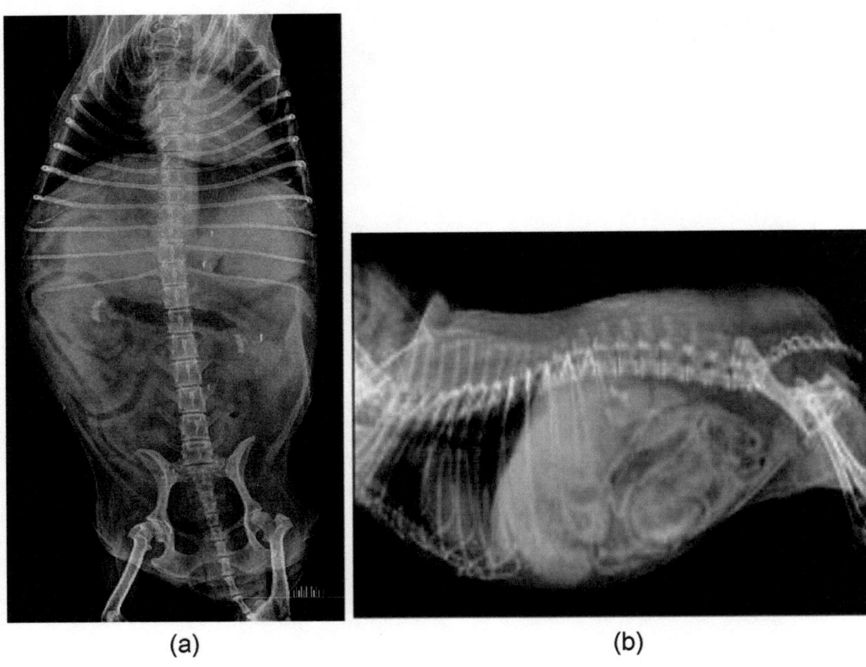

Figure 3.
(a) Right lateral recumbency of abdominal radiography. (b) Dorsal ventral recumbency of abdominal radiography. Two X-ray images were obtained from a Maltese with icterus and abdominal swelling. The liver board line is not obvious, rough, and hepatomegaly.

2.3 Abdominal ultrasound examination and MRI

Abdominal ultrasound examinations are applicable to 94% of patients and allowed determination of an obvious mass in 99% of these patients. The efficiency of the ultrasound method in detecting hepatic mass has been reported previously [1, 16]. In this study, the region containing the pathogenic mass (left, central, or right lobes) was correctly identified. This indicates that ultrasound examination is very sensitive in the identification of a mass as demonstrated in **Figure 4**. Besides the ultrasound examination, the magnetic resonance imaging (MRI) allows one to identify the tumor origin region. A variety of surgical planning modalities benefit

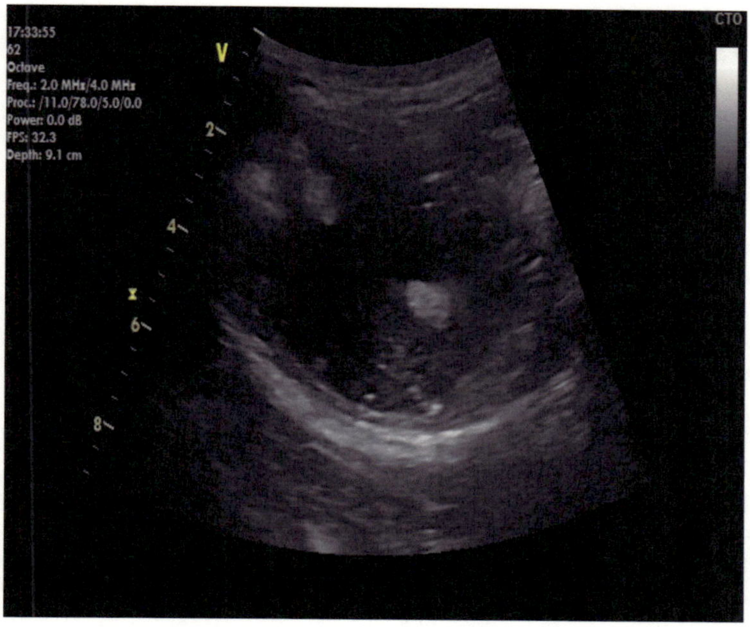

(A)

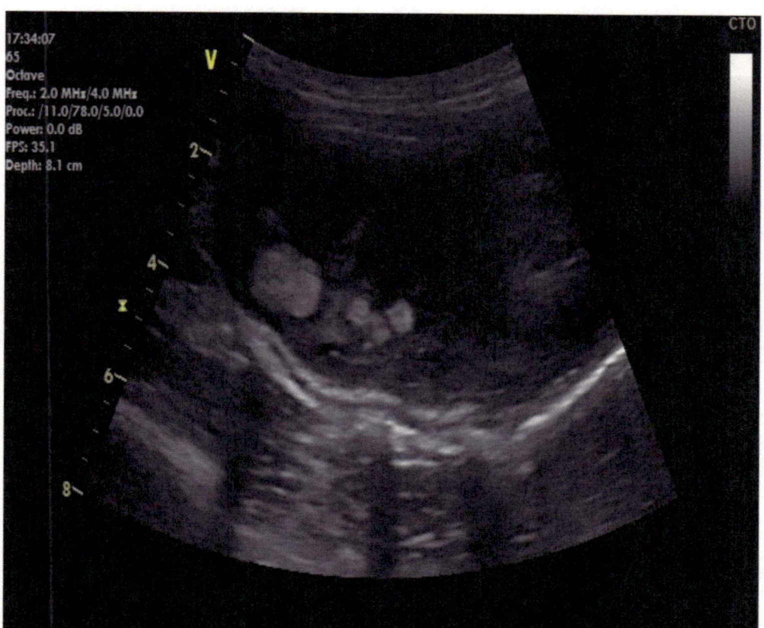

(B)

Figure 4.
(A) Ultrasonography of right medial liver tumor. (B) Ultrasonography of right lateral liver tumor. Survey abdominal sonography 14.5y spayed female cocker spaniel dog before Ga67 examination, several hyperechoic nodules are seen within a large hypoechoic portion in the right medial and lateral hepatic lobe.

from MRI accuracy in the localization of liver tumors; advanced imaging may be needed for precise tumor localization.

2.4 Ga-67 nuclear examination

Canine liver carcinoma cases were analyzed in this study via the nuclear examination technology. We have proposed a preliminary/simplified biokinetic model according to the general-purpose biokinetic model as recommended by the ICRP-30 report, to interpret the empirical data from the clinical examination of a case–control group of hepatic carcinoma dogs [27, 28]. In doing so, every dog was surveyed by a gamma camera/8-slice CT to derive eight complete scanned images within 72 hours. The raw outputs were processed with a self-developed program run in MATLAB to compose a thorough scenario of time-dependent Ga-67 nuclides' intensity changes among multiple compartments (organs or tissues) for dogs that underwent a nuclear examination after being injected the Ga-67-citrate. Each canine patient was administered 22.2 MBq (0.6 mCi) Ga-67 citrate solution via the intravenous injection. The Ga-67 citrate solution was carrier-free with radionuclide and radiochemical purity values exceeding 99.9 and 95.0%, respectively. All radio-pharmaceutical capsules were fabricated by the Syncor International Corporation (USA). The position-sensitive gamma ray emitted from the Ga-67 dose administration for each study object could be robustly assessed and plotted for further analysis. **Figure 5** depicts the eight-compartmental biokinetic model of Ga-67 in liver figuring in the ICRP-30 report, which contains: (1) body fluid, (2) liver, (3) gastrointestinal (GI) tract, (4) kidney, (5) heart, (6) remainder, (7) bladder, and (8) excretion.

Figure 6 illustrates the fragments of scans obtained for a canine patient with liver carcinoma via a gamma camera: (A) The gamma camera/8 slice CT (GE Discovery NM/CT 670) facility indicates that the anesthetized dog was placed between two NM plates for scanning; (B) Close-up view of the anesthetized dog; (C) Particular scans were quoted directly from the E-CAM and implied the raw data

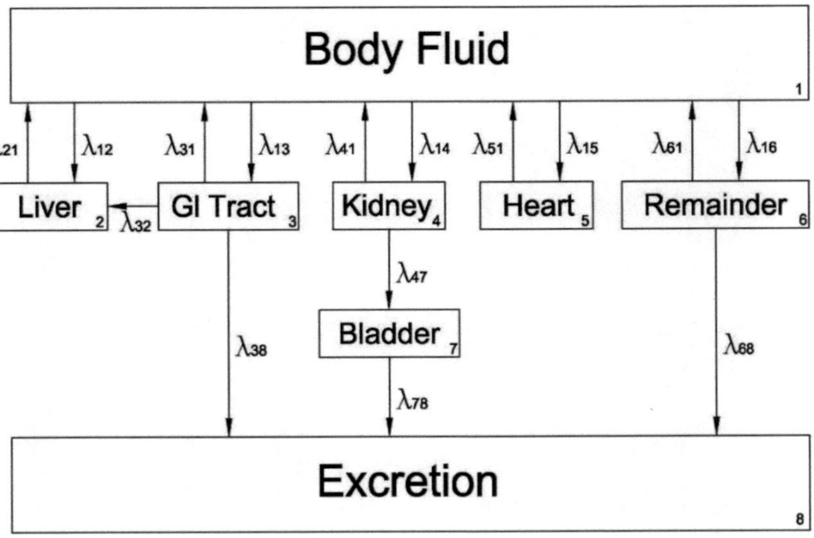

Figure 5.
A simplified biokinetic model of Ga-67 in the liver can be defined by eight major compartments: (1) body fluid, (2) liver, (3) gastrointestinal tract (GI tract), (4) kidney, (5) heart, (6) remainder, (7) bladder and (8) excretion according to ICRP-30 report.

DOI: http://dx.doi.org/10.5772/intechopen.96077

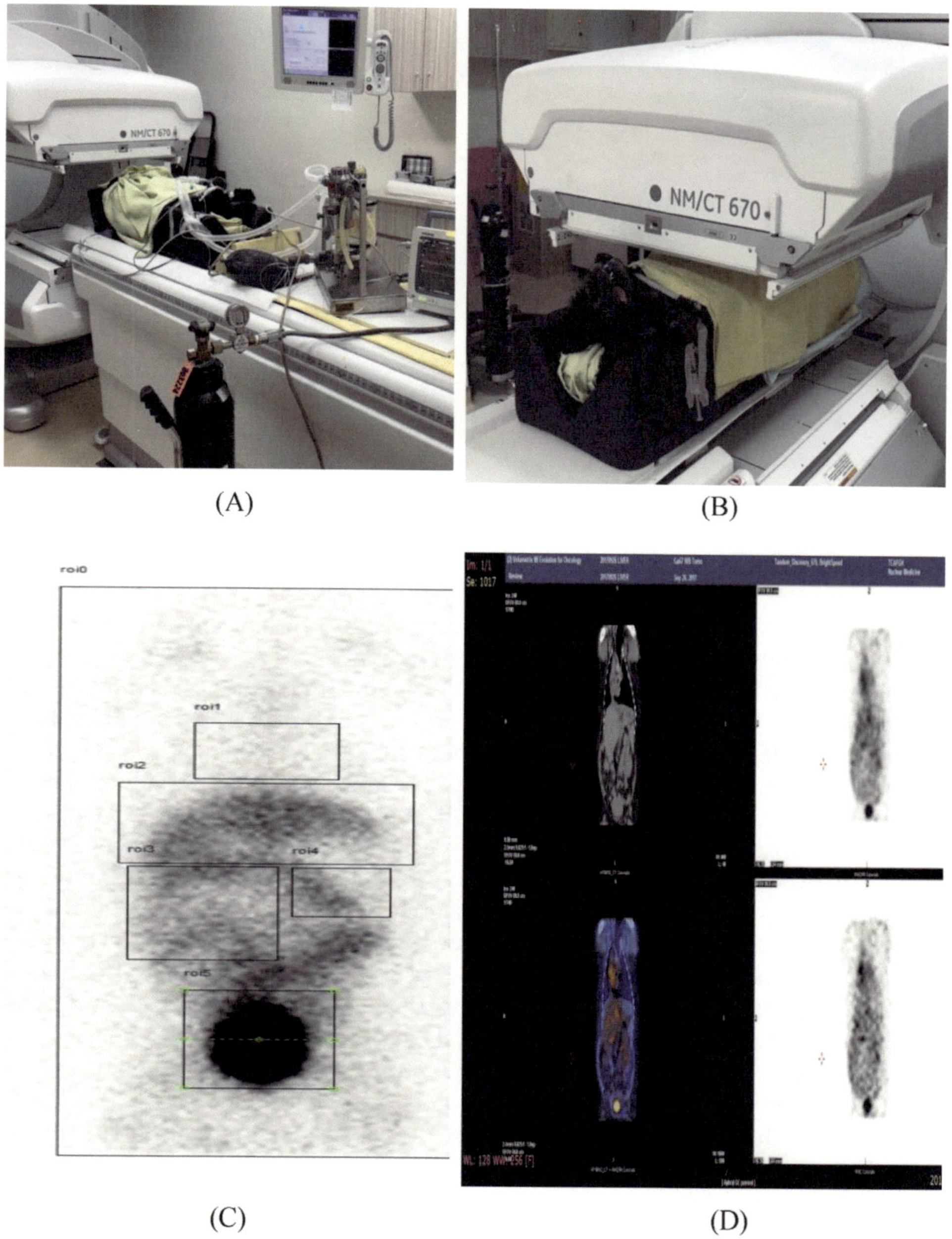

Figure 6.
(A) The gamma camera/8 slice CT (GE discovery NM/CT 670) facility indicates that the anesthetized dog was placed between two NM plates for scanning; (B) close up view of the anesthetized dog; (C) particular scans were quoted directly from the E-CAM and implied the raw data of fifteen-minutes' counting after 72 elapsed hours. Here, the full-size scanning of the dog contained regions of interest (ROI), indicated as follows: ROI0 (whole-body), ROI1 (heart), ROI2 (liver), and ROI3 (GI tract), ROI4 (kidney), and ROI5 (bladder); (D) the fusion plot from gamma camera and CT scanning.

of 15 min counting after 72 elapsed hours. Here, the full-size scanning of the dog contained regions of interest (ROI), indicated as follows: ROI0 (whole-body), ROI1 (heart), ROI2 (liver), and ROI3 (GI tract), ROI4 (kidney), and ROI5 (bladder); (D) The fusion plot from gamma camera and CT scanning. The CT was preset at 130 kV, 150 mA, 0.8 sec., spin width 0.625 mm, spiral speed 3.75 mm/s, and matrix size 512 × 512. The remainder was defined by subtracting all defined compartments (ROI1–5) from the whole body (ROI0).

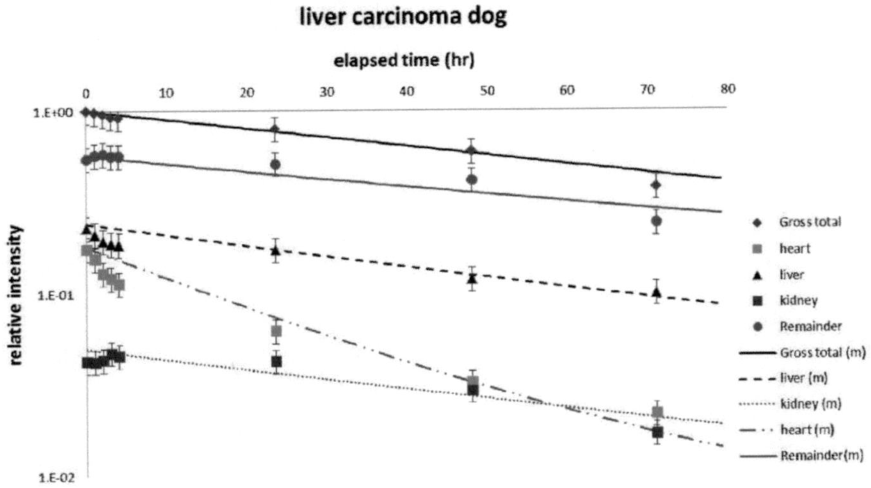

Figure 7.
The time-dependent curves of the Ga-67 concentration among various compartments for a liver carcinoma dog.

Figure 7 shows the time-dependent curves of the Ga-67 concentration among various compartments for the dog with liver carcinoma.

Liver and GI Tract: In contrast to the I-131 thyroid model or GI Tract model for human patients, the model of Ga-67 for liver carcinoma dog was elaborated by the simplification of the general-purpose biokinetic model, according to the ICRP-30 report. As seen in **Figure 5**, a complicated correlation among compartments allows one to use a MATLAB program to optimize the estimation via the empirical data [27]. The feedback path exists between compartments 2–6 and compartment 1. Each compartment has its biological half-life to transfer the Ga-67 radionuclides among compartments, which also confounds the theoretical estimation. Specifically, the liver provides a 22%-contribution of the body fluid (I_{12}, 0.22), which is the second-largest share, whereas the GI Tract has the largest share (I_{13}, 0.33 ~ 0.43) after the Ga-67 administration. However, 60% (I_{21}, 0.60) of the Ga-67 radionuclides returns as feedback to the body fluid with a biological half-life of 15 ~ 40 h. Meanwhile, the remaining nuclides of Ga-67 in GI Tract are transferred to the body fluid (I_{31}), liver (I_{32}), or directly to excretion (I_{38}) with a biological half-life of 20 ~ 600 h. Noteworthy is that the effective half-life is defined as the reciprocal of the sum of reciprocal radiological and biological half-lives ($1/T_{1/2(eff)} = 1/T_{1/2(R)} + 1/T_{1/2(bio)}$). Thus, in practice, either 600 or 20 h of biological half-life still perform as 69 or 16 h of the effective half-life from the continuous gamma camera scanning. Some of Ga-67 radionuclides were found to migrate from the GI Tract to the liver (I_{32}, 0.2 ~ 0.7). Since portal vein circulation provides the blood flow from gastrointestinal section to the liver, the excessive blood pressure will slow the liver's feedback path.

Kidney and bladder. Nearly 20% of Ga-67 nuclides were transferred to the kidney (I_{14}, 0.2), exhibiting nearly no feedback to body fluid (I_{41}, 0.07), in contrast to the bladder (I_{47}, 0.93), and then were fully transferred to excretion (I_{78}, 1.0).

Heart. Only 5–15% of Ga-67 nuclides were transferred to the heart (I_{15}, 0.05–0.15), whereas most of them exhibited an instant feedback to body fluid (I_{51}, 0.99) with a short biological half-time $T_{1/2}$ of 18–20 h. The derived biological half-life of the heart can be treated as a group of cardiac muscles, which provide the blood circulation loop in the whole body, with no apparent Ga-67 nuclides' repository effect.

A long biological half-life of liver for the liver carcinoma dog (40 h vs. 35 h or 15 h) reveals a potential risk of hepatic disorder, whereas the remaining data are barely available to solidify the syndrome of liver carcinoma. The evolution of Ga-67 in the liver carcinoma survey model still plays an essential role in veterinary and human medical domains, since it quantifies the time-dependent concentration of Ga-67 nuclides among compartments of the case–control group and allows one to acquire robust raw data via in-vivo scanning.

3. Global state-of-the-art treatment

3.1 Hepatectomy

The liver surface is convex and it slightly touches the diaphragm. The liver is located on the left side of the caudoventral tract, contacting the stomach, duodenum, pancreas, and right kidney. There are six hepatic lobes: right medial and lateral, left medial and lateral, and quadrate and caudate lobes. The gallbladder is located between the right medial and quadrate lobes. The liver has two so-called afferent (ingoing) blood supplies: the portal system and the arterial system, while the efferent (outgoing) blood flow of the liver circulation is through the hepatic veins. The hepatic lobules, which are the basic functional units of the liver, are cross-sectioned in a hexagonal shape and the portal triads in the periphery. Portal triads consist of the hepatic artery, portal vein, and bile duct. From the anatomic and histologic views, the liver is complicated, and hepatic tissue is friable. Partial lobectomy is difficult and may injure blood vessels and bile ducts in canine patients with bleeding disorders. Many techniques for partial and complete liver lobe resection have been introduced, and numerous stapling instruments have been adopted for both lobectomies. The survey of Liptak et al. [1], which covered 48 dogs with large massive hepatocellular carcinomas during a decade, revealed that their median survival time exceeded 1,460 days after the hepatectomy procedure. However, numerous complications, including ongoing anemia, hepatopathy, ileus, and lack of appetite, are frequently after liver surgery. Therefore, a proper intensive care is recommended to mitigate these complications and minimize the related risks.

3.2 Chemotherapy

The most commonly chemotherapy is administered intravenously. According to clinical chemotherapeutic management of neoplastic cases, a significant advance in veterinary practice is observed. However, a large share of HCC canine patients cannot be cured by chemotherapy and require a further integration of conventional treatment modalities, such as surgery, radiation therapy, and innovative chemotherapy methods. Chemotherapeutic agents are generally administered at the maximum tolerated dose and at the highest dose intensity that is usually used in combination. Four advantages of combination chemotherapy include an increased log-kill, prevention of cancer drug resistance, targeting both dividing and resting cells, and allowing for lower doses with less toxicity [6]. Chemotherapeutic agents damage activated pathogenic cells but also affect normal tissues that divide rapidly and are sensitive to anti-mitotic drugs, such as cells in the bone marrow, digestive tract, and hair follicles. The most common side-effects of chemotherapy are myelosuppression, mucositis, and alopecia. In general, malignant tumors cannot be wholly removed surgically and imply a poor prognosis for canine patients. Palliative chemotherapy and other treatments may be gradually applied to delay the tumor progression. Any further health care should involve close monitoring and minimization of side effects.

3.3 Radiotherapy

As a general rule, surgical resection is considered the best treatment option if a primary tumor can be completely excised. If the region of extensive involvement, normal tissue, or volume of liver tumor make its complete removal problematic, then radiotherapy may be recommended by veterinarians as a palliative treatment of liver tumors. Its effectiveness against the canine liver tumor is limited by the fact that canine patients cannot tolerate cumulative doses exceeding 30 Gy [29]. A share of radiotherapy treatment in US veterinary facilities in 2001 study did not exceed 20% [30], while 92% of facilities in 2010 used the 3D computerized radiotherapy, and 20–100% (with median of 50%) of facilities implemented computer simulation treatment plans [31]. It should be noted the abdominal movement caused by breathing during radiotherapy of liver tumors strongly deteriorates the therapeutic effect, which issue can be resolved for human patients but is hard to control with canine ones.

4. TCVM treatment

4.1 Etiology and pathology

The available methods of animal cancer treatment imply separate or combined application of surgery, radiation therapy, and chemotherapy targeted at the particular cancer cells. However, there is also a general concern about the quality of life of pets/canine patients. This gives much space to the alternative medicine, in particular, the traditional Chinese veterinary medicine (TCVM), which not only focuses on the tumor but also accounts for the overall health condition by regulating the so-called *Yin* and *Yang* constituents. Through the balance of *Yin* and *Yang*, the patients suffering from the disease could also improve their physical health. Tumor, in the TCVM perspective, is the morphological tissue structure change, which imply functional changes of the specific organs or tissues. Those pathological changes of tissues are defined by TCVM as phlegm, toxin, dampness, blood, and stasis. Therefore, the tumor's mechanism can be briefly summarized from the TCVM standpoint as stagnation of blood or (heat-) toxin, accumulative dampness or phlegm, and *Qi (*energy*)* stagnation. The stagnation or lack of a free *Qi*/*Blood* movement results in the formation of pathological tumors in human and animal patients. Those with hepatic carcinomas often have the hormone/gastrointestinal symptoms [Liver (wood) Ke Spleen (Earth)] causing the Spleen *Qi* deficiency.

Spleen *Qi* is responsible for food intake and digestion; this process is called transformation and transportation. Both two functions of the spleen are critical for the production of *Zheng Qi*. *Zheng Qi* deficiency mainly focuses on the root of neoplastic formation. *Zheng Qi* is composed of Nutritive *Qi* and Defensive *Qi* (*Wei Qi*). As *Wei Qi* is a defensive deficiency, several external pathogenic factors (cold, wind, heat, summer heat, dryness, and dampness) cannot be easily detected and expelled from the body. These pathogenic factors will cause the blocking of *Qi* and impede the blood circulation. However, some other factors should be considered: emotional stress, unhealthy diet/lifestyle, and the environment. For example, negative emotional stress, inappropriate diet, and too humid environment are considered by TCVM as induction factors that trigger a liver tumor. Both internal and external factors may contribute to phlegm, *Qi,* and *Blood* stagnation, blood stasis, and ultimately lead to neoplasia [12, 19, 21, 22].

4.2 Pattern differentiation and treatment

The TCVM has been used as an alternative treatment for years in Asia. However, its current applications to domestic animals/pets suffering from carcinomas are based on individual expert opinions, while there are no outlined veterinary treatment strategies and guidelines for clinical practice in this field. The most lucrative concept accepted nowadays is a comprehensive combination of global/Western and TCVM components, the latter being aimed at adjunct therapy and recurrence prevention. Adjunct therapy should reduce the side-effects of chemotherapy, radiation therapy, and surgery. Based on the pattern differentiation, it is essential to treat the liver tumor using TCVM drugs and acupuncture techniques capable of regulating *Qi*, nourishing blood, strengthening the body and organs, and improving the resistance to pathogenic factors. The TCVM is likely to improve the canine patient's general conditions, remove the disease/pathogen, inhibit oncogenesis, alleviate side-effects, and improve the survival rate as shown in **Figure 8**.

The TCVM treatment is usually provided to canine patients undergoing a surgical treatment or after radiotherapy/chemotherapy. In most of these patients, tumors could not be eliminated entirely, and the adjunct treatment should improve their

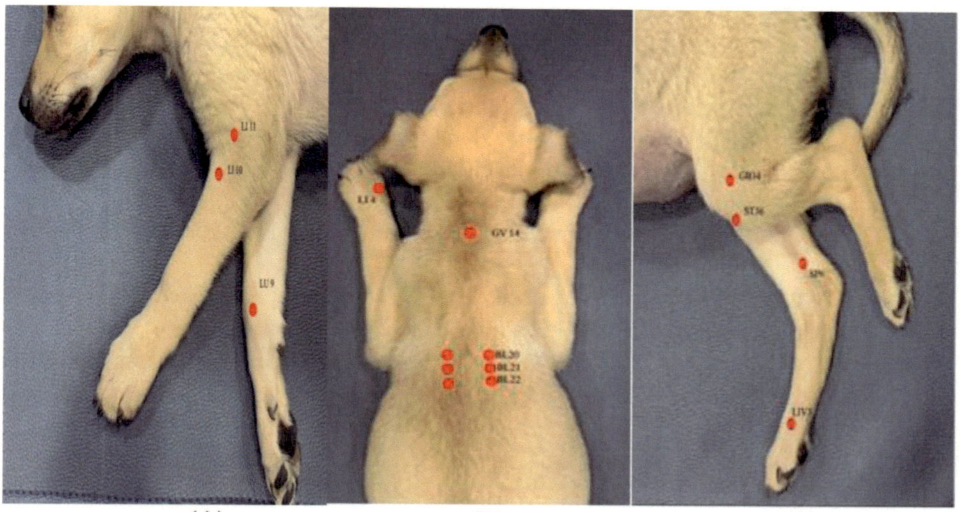

(A) (B) (C)

Figure 8.
Acupuncture points. (A) LU 9: In the depression distal palmar to the most medial prominence of the radial styloid process, overlying the radiocarpal joint, medial to the radial artery, and the tendon of the flexor carpi radialis muscle. LI 10: In the depression in the muscular groove between the extensor carpi radialis and the common digital extensor mm. Of the forelimb, two cun distal to the transverse cubital crease. This is most evident when the elbow is flexed.LI 11: In the depression in the transverse cubital crease, just cranial to the lateral epicondyle of the humerus, between the extensor carpi radialis and common digital extensor mm. This point is easily palpated when the elbow is flexed. (B) LI 4: In the depression between the 2nd and 3rd metacarpal bones, approximately in the middle of the 2nd metacarpal bone. GV 14: In the depression on the midline between the dorsal spinous processes of the 7th cervical and 1st thoracic vertebrae. BL 20: In the depression, 1.5 cun lateral to the caudal border of the spinous process of the 12th thoracic vertebra. BL21: In the depression, 1.5 cun lateral to the caudal border of the spinous process of the 13th thoracic vertebra. BL 20: In the depression, 1.5 cun lateral to the caudal border of the spinous process of the 12th thoracic vertebra. BL22: In the depression, 1.5 cun lateral to the caudal border of the spinous process of the 1st lumbar vertebra. (C) LIV 3: In the depression on the dorsum of the rear foot, between the 2nd and 3rd metatarsal bones, at the level of the junction of their heads and shaft, just proximal to their associated metatarsophalangeal joints. SP 6: In the depression 3 cun proximal to the tip of the tibia's medial malleolus, on the caudal border of the tibia. GB34: In the depression cranial and distal to the head of the fibula. ST36: In the depression, just lateral to the distal aspect of the cranial border of the tibial tuberosity (tibial crest), approximately in the middle of the cranial tibialis muscle.

quality of life, maintain their physical condition, and prolong survival time. TCVM has shown significant efficacy in the symptomatic treatment of canine patients suffering from the deficiency of vital *Qi*, leading to physical pain, fever, anorexia, nausea, gastrointestinal problems, fatigue, and constipation. Although liver heat and dampness are considered as the most probable causes for the formation of liver carcinoma and viral hepatitis in human patients, the animal/canine ones are less prone to viral pathogens. From the TCVM perspective, there are two patterns controlling liver carcinomas in small animals/dogs.

4.2.1 The first pattern: "blood stasis with Qi deficiency"

The *Qi deficiency* is more specifically related to "*Spleen Qi deficiency*", leading to "Blood Stasis". The main TCVM principles are to improve blood circulation, nourish the blood, resolve the stagnation or accumulation, tonify the spleen, and relieve pain. Refer to **Table 4**; acupuncture on specific points can improve and enhance the symptoms described earlier. *Xiao Yao San*, which is compounded by various herbs, is a transitional Chinese medicine described in the "Formulary of the Tai Ping Welfare Dispensary Bureau' Collections of Medicinal Formulations" compiled by Chen Shi-wen et al. in 1151. A modified herb formula of *Nei Xiao Wan*, called "Stasis Breaker", has been introduced by *Xie* et al. specifically for animals [19, 32] to reduce phlegms and stasis, clear toxin substances, and promote *Qi* and *Blood* circulation. The significant effect of "Stasis Breaker" breaks the blood stasis, softening the hard nodes and tumors. Furthermore, some studies prove that *Bai Hua She She Cao* and *Ban Zhi Lian* can inhibit cell mutation, tumor growth and clear the "heat-toxin" [19, 33, 34].

4.2.2 The second pattern of liver carcinoma: "blood stasis with Yin deficiency"

Heat and liver stagnation are supposed by TCVM to result in *Qi* and *Blood* stagnation. The consumption of fluid will injure the *Yin* in the *Middle Burn*. Once *Liver Yin* and *Kidney Yin* are injured, the tumor will be gradually formed. This treatment pattern focuses on nourishing *Yin* and *Blood*, resolving pain and stagnation, tonifying *Qi*, and resolving tumor. The recommended acupuncture points are shown in **Table 4**. [35]. *Yi Guan Jain*, a herbal medicine, described in the "Supplement to the Classified Case Records of Famous Physicians" by *Wei Zhi-Xiu* in 1770. It is intended to tonify the *Liver's* and *Kidney's Yin* and clear the "false heat". The combination of *Yi Guan Jain* and "Stasis breaker" effectively improves the immune system

Symptom	Acupuncture points
General *Qi* deficiency	LI -10, ST-36, CV-17, LU-9
Spleen deficiency/eliminate phlegm	BL-20, BL-21, ST-36, ST-40
Immunization	GV-14, LI-4, LI-10, LI-11
Smooth *Qi* and relieve pain	LIV-3, LIV-4, GB-34, GB-41
Anorexia	CV-12, Shan-gen
Vomiting	PC-6, GB-34, CV-12
Diarrhea	GV-1, SP-6
Ascites	SP-6, SP-9, BL-22,

Table 4.
TCVM acupuncture points for tumor.

performance, inhibiting tumor growth, mutation, and metastasis. This combination can also mitigate chemotherapy/radiotherapy side-effects and improve life quality [12, 19, 21, 22].

5. Conclusions

The global veterinary medicine is widely used to eliminate canine carcinogens by substantiated combinations of drug administration, surgery, chemotherapy, and radiotherapy treatments. However, their side-effects remain a challenging issue faced by clinical veterinarians and dog owners. In this respect, traditional Chinese veterinary medicine (TCVM), including acupuncture, herbs, food therapy, and massage, is considered a lucrative integrated treatment system. Using the pattern differentiation, as well as the proposed biokinetic Ga-67 model of canine liver carcinoma, the robust HCC diagnosis of canine patents can be obtained. A further application of global state-of-the-art and/or TCVM-based therapies can enhance the immune system, speed up the recovery, relieve pain syndrome, reduce chemotherapy toxicities, and improve quality of life of canine patients. Thus, the global integrative oncology comprehensively combines regulated clinical treatments with complementary and alternative medicine (TCVM in particular), yielding the synergistic curing effect.

Author details

Chih-Feng Chen[1,2,3], Shao-Wen Chiu[1,2,4], Lung-Kwang Pan[1]* and Chao-Hsun Chuang[1,2,5]*

1 Graduate Institute of Radiological Science, Central Taiwan University of Science and Technology, Takun, Taichung, Taiwan, ROC

2 Bachelor Degree Program in Pet Healthcare, Yuan-Pei University of Medical Technology, Xiang-shan, Hsin-chu, Taiwan, ROC

3 Yong-Shin Animal Hospital, Takun, Taichung, Taiwan, ROC

4 Healthcare Technology Business Division, Healthcare Department, International Integrated Systems Inc., Taipei, Taiwan, ROC

5 Chyun-Min Animal Hospital, Chunan, Taiwan, ROC

*Address all correspondence to: lkpan@ctust.edu.tw and petslovers@yahoo.com.tw

References

[1] Liptak JM, Dernel WS, Monett E, Powers B, Massive hepatocellular carcinoma in dogs: 48 cases (1992-2002), J Am Vet Med Assoc. 2004; 225(8):1225-30. DOI: 10.2460/javma.2004.225.1225.

[2] Selmic LE, Hepatobiliary Neoplasia, Vet Clin North Am Small Anim Pract, 2017; 47(3):725-735. DOI: 10.1016/j.cvsm.2016.11.016. Epub 2017 Jan 5.

[3] Balkman C, Hepatobiliary neoplasia in dogs and cats, Vet Clin North Am Small Anim Pract. 2009; 39(3):617-25.

[4] Magne ML, Withrow SJ, Hepatic neoplasia, Vet Clin North Am Small Anim Pract, 1985; 15(1):243-56.

[5] Thompson MB, Chappell JD, Kunze DJ, et al., Bile acid profile in a dog with cholangiocarcinoma, Vet Pathol. 1989; 26(1):75-8.

[6] https://veteriankey.com/chemotherapy/, chapter 44: Chemotherapy

[7] Orna K, Kenneth MR, John MG, et al., Hepatotoxicity associated with CCNU (lomustine) chemotherapy in dogs, J Vet Intern Med. Jan-Feb 2004; 18(1):75-80. doi:10.1892/0891-6640.

[8] Webster CRL, Center SA, Cullen JM, et al., ACVIM consensus statement on the diagnosis and treatment of chronic hepatitis in dogs, J Vet Intern Med. 2019; 33(3):1173-1200. Epub 2019 Mar 7.

[9] Patnaik AK, Hurvitz AI, Lieberman PH. Canine hepatic neoplasms: A clinicopathologic study. Vet Pathol. 1980; 17:553-564.

[10] Monnet E, Kudnig S. Alimentary tract, liver, and pancreas. In: Slatter DH, editor. Textbook of Small Animal Surgery. Philadelphia, Pennsylvania: WB Saunders; 1993; 2378-2382.

[11] Liptak JM. Hepatobiliary tumors. In: Withrow SJ, Vail DM, editors. Withrow & MacEwen's Small Animal Clinical Oncology. St. Louis, Missouri: Elsevier; 2007; 483-490.

[12] Guan YS, He Q. Liver cancer: Zheng Classification of Qi stagnation and blood stasis. Pharmacology & Pharmacy, 2014, 5, 75-82, http://dx.doi.org/10.4236/ pp.2014.51012

[13] Liptak JM, Dernell WS, Withrow SJ. Liver tumors in cats and dogs. Compend Contin Educ Pract Vet. 2004; 26:50-56.

[14] Guilford WG, Strombeck DR. Hepatic neoplasms. In: Guilford WG, Strombeck DR, Williams DA, editors. Strombeck's Small Animal Gastroenterology. 3rd ed. Philadelphia: Pennsylvania: Saunders; 1996. 847-859.

[15] Patnaik AK, Hurvitz AI, Lieberman PH, et al., Canine hepatocellular carcinoma. Vet Pathol. 1981; 18:427-438.

[16] Kosovsky JE, Manfra-Marretta S, Matthiesen DT, et al., Results of partial hepatectomy in 18 dogs with hepatocellular carcinoma. J Am Vet Med Assoc. 1989; 25:203-206.

[17] Mori T, Ito Y, Kawabe M, et al., Three-dimensional conformal radiation therapy for inoperable massive hepatocellular carcinoma in six dogs, https://doi.org/10.1111/jsap.12352

[18] Dobos GJ, Kirschbaum B, Choi KE, The Western model of integrative oncology: the contribution of Chinese medicine, Chin J Integr Med. 2012; 18(9):643-651. DOI: 10.1007/s11655-012-1200-1. Epub 2012 Aug 31.

[19] Xie HS, Preast V, Traditional Chinese Veterinary Medicine: Fundamental Principles, Chi Intitute Press; 2nd edition (January 1, 2013), ISBN-10: 1934786411.

[20] Marconato L, Sabattini S, Marisi G, et al., Sorafenib for the treatment of unresectable hepatocellular carcinoma: preliminary toxicity and activity data in dogs, Cancers 2020; 12:1272; DOI:10.3390/cancers12051272

[21] Leifer CE, Peterson ME, Matus RE, et al. Hypoglycemia associated with nonislet cell tumor in 13 dogs. J Am Vet Med Assoc. 1985; 186:53-55.

[22] Strombeck DR, Krum S, Meyer D, et al. Hypoglycemia and hypoinsulinemia associated with hepatoma in a dog. J Am Vet Med Assoc. 1976; 169:811-812.

[23] Quigley KA, Jackson ML, Haines DM. Hyperlipasemia in 6 dogs with pancreatic or hepatic neoplasia: evidence for tumor lipase production. Vet Clin Pathol. 2001; 30:114-120.

[24] Lowseth LA, Gillett NA, Chang IY, et al. Detection of serum alpha-fetoprotein in dogs with hepatic tumors. J Am Vet Med Assoc. 1991; 199:735-741.

[25] Yamada T, Fujita M, Kitao S, et al. Serum alpha-fetoprotein values in dogs with various hepatic diseases. J Vet Med Sci. 1999; 61:657-659.

[26] Hammer AS, Sikkema DA. Hepatobiliary tumors. In: Withrow SJ, MacEwen EG, editors: *Small Animal Clinical Oncology*, Ed 3, Philadelphia, 2001, WB Saunder.

[27] Chuang CH, Chen CF, Tseng NC, et al. The first attempt of the biokinetic Ga-67 model application to canine liver carcinoma: case-control study via in-vivo gamma camera/8-slice CT technique. Journal of Mechanics in Medicine and Biology, 2018; 18(7):1840019 https://doi.org/10.1142/S0219519418400225.

[28] Chen CF, Chuang CH, Tang PC, et al. Biokinetic model development of radioiodine-131 via in-vivo gamma camera/8-slice CT technique: Case-control study of feline hyperthyroidism. Journal of Mechanics in Medicine and Biology, 2018; 18(8):1840035 https://doi.org/10.1142/S0219519418400353

[29] Bartlett DL, Carr BI, Marsh JW. Cancer of the liver. In DeVita VT, Hellman S, Rosenberg SA, editor: *Cancer: Principles and Practice of Oncology*, Philadelphia, 2005, Lippincott Williams & Wilkins.

[30] McEntee MC. A survey of veterinary radiation facilities in the United States during 2001. Vet Radiol Ultrasound. 2004; 45:476-479. [PubMed: 15487574]

[31] Farrelly J, McEntee MC. A survey of veterinary radiation facilities in 2010. Vet Radiol Ultrasound. 2014; 55:638-643. [PubMed: 24798372]

[32] Xie HS, Lindsey W, Cheryl LC, et al., Practical Guide to Traditional Chinese Veterinary Medicine, Vol. 2: Small Animal Practice, August 1st, 2014, EdS, DACVIM-Neurology

[33] Patnaik AK, Lieberman PH, Erlandson RA, et al. Hepatobiliary neuroendocrine carcinoma in cats: a clinicopathologic, immunohistochemical, and ultrastructural study of 17 cases. Vet Pathol. 2005; 42:331-337.

[34] Patnaik AK, Newman SJ, Scase T, et al. Canine hepatic neuroendocrine carcinoma: an immunohistochemical and electron microscopic study. Vet Pathol. 2005; 42:140-146.

[35] Xie HS, Preast V, Xie's Veterinary Acupuncture, 1st Edition, Blackwell Publishing, 2017, ISBN-13: 978-0813812472